Comets in Astrology

Περισκοπεῖν ἄστρων δρόμους ὥσπερ συμπεριθέοντα
"Observe the movements of the stars as if you circled with them."

MARCUS AURELIUS
26 APRIL 121 CE – 17 MARCH 180 CE

Comets in Astrology

Rod Chang

THE WESSEX ASTROLOGER

Published in 2025 by
The Wessex Astrologer Ltd
PO Box 9307
Swanage
BH19 9BF

For a full list of our titles go to www.wessexastrologer.com

© Rod Chang 2025

Rod Chang asserts his moral right to be recognised as the author of this work

ISBN 9781916625303

Cover design by Fiona Bowring at Bowring Creative
Front cover image: Comet C/2023 A3 Tsuchinshan-ATLAS
taken by the author on Mount Teide, Tenerife

Typeset by Kevin Moore

A catalogue record for this book is available at The British Library

No part of this book may be reproduced or used in any form or by any means without the written permission of the publisher. A reviewer may quote brief passages.

Contents

Acknowledgements	vi
Preface	vii
Foreword	viii
Chapter 1: Comets in Astronomy	1
Chapter 2: A Brief History of Comet Observation	6
Chapter 3: Ancient Astrologers on Comets	11
Chapter 4: Mythology	20
Chapter 5: A Modern Perspective on Comets	22
Chapter 6: Comets and Natal Charts	30
Chapter 7: How to find comets using the JPL Horizons system	50
Chapter 8: Comet Case Studies	54
Halley's Comet 1066	55
The Great Comet of 1577	60
Great Comet of 1677	64
Great Comet of 1844	66
Great Comet of 1901	70
Great Comets of 1910	75
Comet Ikeya–Seki	86
Comet Halley	93
Comet Hale–Bopp	107
Comet McNaught	119
Comet Panstarrs	123
Comet Wirtanen	133
Comet Neowise	140
Comet Leonard	152
Comets of 2024	170
Conclusion	185
Notes & References	186

Acknowledgements

I would like to thank all my teachers, mentors, colleagues and friends who have helped me on my astrological journey. Carole Taylor, my first teacher at the Faculty of Astrological Studies, who encouraged me to study comets and helped me to publish my first article on the subject in the *Astrological Journal* in 2014. The teachings of Melanie Reinhart, Bernadette Brady and Sharon Knight have had an important influence on how I do research. I would also like to thank my friends Jupiter, Brian, Nusse, Mandi Lockley, Marcos Patchett, Christian König, Tania Daniels, Giulio Pellegrini, Lynn Bell, Darby Costello, Lianne McCafferty, Luiza Azancot, Christeen Skinner, Geraldine Williams, Wendy Stacey, Kenan Yasin and Alejo Lopez who have always encouraged me and helped me in my astrological journey. I also give special thanks to my dear sister Yvetta who has always supported me, and I could not have finished this book without Peter's help in editing, proofreading and support.

I would like to give special thanks to Margaret Cahill and the team at The Wessex Astrologer for publishing this book. Margaret publishes many excellent books and by doing so makes an important contribution to our astrological community. I am honoured to be part of her family.

Preface

Look at the night sky

When the comet Panstarrs (C/2011 L4) appeared in the sky back in 2013, I was approached by a reporter asking what astrological influence it would have. The question made me realise how little I knew about comets, and prompted me to start studying the phenomenon, while at the same time reawakening my relationship with the night sky and my love of astrophotography. The night sky gives us so many celestial objects to observe, discover and learn about. I feel that sometimes we should just put down our charts, papers, books and apps, go outside and spend more time looking up at the actual heavens above us.

For the last eleven years I have been studying how ancient astrologers interpreted comets and have kept asking myself what other meanings they might hold. Now, after collecting more and more data and learning how to generate a comet ephemeris, I believe I have a clearer picture of their astrological meaning. From a harmonious universe perspective, comets are seen as disruptors, irregular objects in the solar system. As astrologers we have difficulty calculating their position or even finding them in astrological software. I hope this book will encourage the reader to take more of an interest in these extraordinary celestial objects that have perplexed, fascinated and even frightened stargazers since time immemorial.

Best wishes,

Rod Chang
Summer 2025

Foreword

Humankind has always looked to the starry sky for guidance. The Gospel of Luke in the New Testament states that 'great signs shall there be from heaven'. The defeat of Hannibal in 20BC was apparently foretold by the arrival of a comet. Fast forward a few centuries and we have evidence that astrologers of the 16th century took note of the correlation between visible-to-the-naked-eye comets and noteworthy events.

For thousands of years attention has been given to the potential impact of comets on life on Earth. Toward the end of the 19th Century, several books and articles covered the fascinating subject. A few articles appeared in the first half of the 20th century, but there was a lull for some decades before a few works appeared at the turn of the millennium. Just as news of visible comets capture attention, so too do books on the subject suddenly appear as if from nowhere. This important work is to be welcomed and will surely be a valuable addition to your library.

The care and attention given in this work differs from earlier works in that Rod Chang offers many examples complete with zodiacal positions. From considering the chart for a comet's discovery to the comet's course through particular areas of the zodiac, he breaks new ground.

It is exceedingly well-researched and thought-provoking. The thorough examination of both the colour and shape of comets and how these factors confirm ideas and concepts that have captivated mankind's imagination for centuries, makes clear that Rod is a master in this under-explored area.

It is curious that although the many astrology software programs allow us to add numerous asteroids, fixed stars, dwarf and trans-Neptunian planet positions to our charts, they do not – as yet - offer the insertion of comet positions. Those reading this book will surely clamour for this option to be added.

We recognise that our solar system is in constant change. In recent times the amount of natrium in the Moon's atmosphere has increased, Venus, Uranus and Neptune have each appeared brighter and solar activity is behaving in unexpected ways, which is clearly having an impact on the entire solar system. Even the icecaps on Mars are melting.

We know all this thanks to the exemplary work of today's astronomers and the technicians who have advanced the capabilities of Galileo's telescope in their reach to know more about our universe. We owe gratitude to these astronomers and mathematicians who identify and plot the course of comets. We should also revere Rod Chang for both identifying which comets are likely to have greatest impact on life on Earth and for suggesting ways in which to interpret their influence.

Astrologers of the early 21st century are exceedingly challenged. Not only are they learning more about the physical attributes of each of the known planets and how these are altered by solar activity, but they have also been incorporating and interpreting the position of major asteroids into their charts. More recently, they have embraced knowledge of the planetary bodies orbiting in the Kuiper belt at the edge of the solar system. It can all seem too, too much and it would be understandable if astrologers concluded that their charts are busy enough already.

The effort of placing comets (most likely by hand) in their charts might seem a step too far. However, on reading this work they will surely conclude the effort to be worthwhile.

Expect to be captivated and enthralled. Rod Chang writes clearly and might soon be known as a 'Master Cometeer'.

Christeen Skinner
June 2025

Chapter 1

Comets in Astronomy

In astronomy, a comet is an object in space which has an elongated elliptical orbit. Some comets are visible in the sky at regular intervals within a relatively short period of time. For example, the famous Halley's Comet is visible from Earth every 75 to 76 years. Others have an extremely long orbital period, visiting the Sun once in hundreds of thousands of years.[1]

In Western languages like French (comète), German (Komet) and English (comet), the word is derived from the Latin 'cometes'. This is the Latin translation of the Greek κομήτης (komētēs) whose literal meaning is 'long hair'. The singular Greek word κόμη (komē) originally meant 'the hair on one's head'.[2]

In the Chinese language, comets are called 'Hui Xin'. In Shuo-wen 《說文》, 'Hui' '篲' means "bamboo brooms".[3] So the ancient Chinese often referred to comets as 'broom stars', because the tail of a comet looks like a bamboo broom.

Types of comets and where they come from

Most comets come from pretty far away: the asteroid belt between Mars and Jupiter, the Kuiper Belt, and the Oort Cloud.

The Kuiper Belt extends from Neptune's orbit of around 30 astronomical units (AU) to about 50 AU from the Sun. The Kuiper Belt is mainly composed of icy bodies, and the comets coming from there are mostly 'short-period' comets, such as Halley's Comet.[4]

Chiron was initially called an asteroid and classified as a minor planet with the designation '2060 Chiron'. However, in 1989 it was found to exhibit behaviour typical of a comet. Chiron is now classified as both a minor planet and a comet and is accordingly also known by the cometary designation 95P/Chiron.[5]

The Oort Cloud is at the gravitational edge of the Sun (about 50,000 to 200,000 AU) and is also made up of icy bodies. Most of the comets from here are 'long-period' comets, with orbits lasting hundreds of thousands of years.[6]

A comet with an orbit around the Sun that has an eccentricity greater than 1 is called a 'hyperbolic comet'. Some comets may be put into hyperbolic orbits after passing close to Jupiter. This happened with Lexell's Comet in 1779. Comets on such trajectories are lost to the solar system. Some comets making their first visit to the inner solar system from the Oort Cloud may follow hyperbolic trajectories, never to return.[7] 2024's comet C/2023 A3 (Tsuchinshan–ATLAS) could be hyperbolic and it's possible some hyperbolic comets come from outside our solar system.

'Main-belt' comets exist within the orbit of Jupiter and have only been discovered in recent years. They are also called active asteroids. They have the same orbits as asteroids, and are smaller than other types of comet. At perihelion, they will appear like comets with tails. P/2010 A2 (LINEAR) was initially thought to be a normal comet, but it was later found to have an orbit in the main asteroid belt and was determined to be a main-belt comet.[8]

Nucleus and Coma

Comets come from distant, cold and dark space. As the Sun's gravity pulls them closer, they enter our field of vision. They rise in temperature as they approach the Sun causing tails to form as radiation. Solar winds evaporate frozen gases and ice trapped inside them. The resulting illuminated tail is a comet's trademark 'coma'.[9] Many comets will burn up and then split into pieces or even disappear when they pass by the Sun, and even the Great Comets can split up into several smaller comets, some of which will disappear after they pass the perihelion.

Comet nuclei are composed of ice, dust and frozen gases, including carbon monoxide, methane and ammonia.[10]

The surface of a comet's core is made up of rock and dry dust. Beneath the rock lies a layer of ice. This ice and dust will become the coma as the comet approaches the Sun. However the deeper nuclei may also contain a variety of organic compounds, including methanol, hydrogen cyanide, formaldehyde, ethanol, ethane, and more complex molecules such as long-chain hydrocarbons and amino acids. A report based on NASA studies of meteorites found on Earth shows that DNA and RNA components (adenine, guanine and related organic materials) may have formed on asteroids and comets.[11]

When comets come close to the Sun, both their core and tail are illuminated, and that is when we can observe them. The cores of many comets are not large, but their tails may be thousands or millions of kilometres long. When a comet comes as close as the orbit of Mars, the solar wind may blow and expand its tail into beautiful shapes. Each comet's tail of gas and dust is unique and points in a slightly different direction.

The advancement of astronomical observation technology has revealed that many comets have multiple tails. Those containing dust are called dust tails. There are also ion tails, where it is thought gases in the cometary comae are ionised by solar ultraviolet photons. These ionised gases are then affected by the magnetic forces of the solar wind. The ions are swept out of the coma and into a long, distinctive ion tail. This tail often appears blue to the human eye because the most common ion, CO+, scatters blue light more than red. The ion tail may be bent into ropes, knots and streamers that point away from the Sun by changes in the magnetic field of the fast-moving solar wind, which can make it look different to the dust tail.[12]

Brightness: How bright could a comet be?

A comet does not emit light. Like other celestial bodies in our solar system, it can only reflect the Sun's light. The size of the core is therefore a key factor in how bright it will appear in the sky. It is a well-established fact that comets with large cores become brighter when they approach the Sun. The comet's surface composition can also affect its brightness. Astronomers have found that comets reflect only 3-4% of the Sun's light. The vast majority of comets are very small and therefore cannot be seen when they are far away. Even when they are close to the Sun, they are only a small point of light that is difficult to spot. The brightness of a comet is also affected by what happens to its nucleus.[13] Some comets will suddenly increase their brightness, as Comet 12P (Pons-Brooks) did in July 2023. An eruption caused its brightness to increase by about 100 times, jumping from magnitude 17 to 12 overnight.[14] Obviously it is when the comet comes closest to the Sun and the Earth that we can observe it most easily.

The well-known comet Hale–Bopp attracted a great deal of attention when it appeared on 22 March 1997. Hale–Bopp had its closest approach to Earth at a distance of 1.315 AU (196 million km), which is not very close compared to other comets. However its core was huge and active, with an average diameter of 37 miles (60 km), making this a Great Comet.[15]

In contrast, Comet Neowise in July 2020 had a diameter of only 3 miles (5 km), yet its closest distance to Earth was 0.7 AU (104 million km). It looked like a little hair in the night sky, and was hard to detect.

Comet break up and meteor showers

When observing comets, it is often said that they may break up when passing perihelion. This is what happened with the Great Comets West and Ikeya–Seki. In fact, comets with fragile cores may split apart due to thermal stress or internal gas pressure at any time in their orbit. They could also be split apart by impact.[16]

The larger dust particles are left behind along the comet's orbital path as it moves across the sky, while the smaller particles are pushed away from the Sun into the comet's tail by the gentle pressure of solar winds. When the Earth's orbit passes through the debris of these comets, it causes meteor showers. These include the Eta Aquarius meteor showers between late April and May each year and the Orionids between October and November. Both are caused by debris from Halley's Comet. The well-known Perseids, which appear every July and August, are the result of debris from the Great Comet of 1862.[17]

Comets are unpredictable and can be disappointing

When people hear that a comet will soon appear in the sky, some might expect to see something the size of the Moon. In fact, there is no precise scientific definition of a Great Comet, but if it appears large and bright to the naked eye then that is what it could be. Their appearance is unpredictable. Some people say they occur once every ten years on average. The last Great Comet was Comet McNaught in 2007 until Comet Tsuchinshan appeared in 2024.

Comets are like rebels in the solar system. We cannot predict their whereabouts with certainty or how often they will appear. We can say with some confidence that there will be a chance to see a short-period comet every 2-3 years, but you simply cannot see these small comets without a telescope. In fact many of the comets reported in the media just look like a tiny dot of light to the naked eye.

Comets' unpredictability and weird behaviour must be added to their interpretation by astrologers. They hide at the edge of the solar system and suddenly appear, taking us by surprise. Comet Nishimura (C/2023 P1), which was suddenly discovered in August 2023, had been hiding behind the Sun's rays. Had a photographer not been observing carefully, it would never have been noticed.

Sometimes comets let us down. For example, Halley's Comet in 1986 failed to display a beautiful tail, as all amateur astronomers had expected. One of the most famous examples was Comet Kohoutek (C/1973 E1). This long-period comet with a 150,000 year orbit appeared in March 1973, but it disappointed everyone. Its core measured just 2.6 miles (4.2 km), its perihelion was only 0.14 AU (20 million km) from the Sun, and its closest distance to the Earth was only 0.8 AU. Though it was visible to the naked eye, and made people believe they were going to see a Great Comet, it could not be observed for long, and was not as bright as people had expected, making it a most disappointing sight. American astronomer Whipple was unequivocal in his assessment: "If you want to have a safe gamble, bet on a horse – not a comet."[18] Before its appearance, the media called Comet Kohoutek "the comet of the century". The media never changes, always exaggerating stories to boost sales.

Chapter 2

A Brief History of Comet Observation

The earliest confirmed observation of a comet was in China. The Spring and Autumn Annals record that "in the autumn of the seventh lunar month, a comet entered the Big Dipper (秋七月, 有星孛入於北斗) in 613 BC.[19]

In the chapter 'Basic Annals' of Qin Shi Huang (始皇本紀) of *The Records of the Grand Historian* (史記), it is recorded that a comet appeared in 240 BC:

> "In the 7th year of Qin Shi Huang, a Comet first appeared in the East, and can be observed in the North direction, in the 5th Month you can see it in the West, it lasted 16 days (始皇七年, 彗星先出東方, 見北方; 五月見西方, 十六日)."[20]

Modern scientists have determined that these were observations of Halley's Comet. It has also been speculated that the Chinese may have observed Halley's Comet in 1059 BCE, though others suggest this sighting was actually only calculated later after it had appeared.

The ancient Greek scholars, including Aristotle, all considered comets to be abnormal phenomena, but different schools had different opinions about what they were. Most scholars believed comets were simply illusions resulting from disturbances in the upper layers of the atmosphere. Their brightness in the sky led them to believe comets were caused by extremely hot, arid and windy weather and correlated them with hurricanes, droughts and rainstorms, describing them as "big fire" in the sky. Another early astrological study suggested that comets appeared when several planets formed a conjunction. These two views led to much argument and debate in scholarly circles.[21]

There were hurricanes when comets were observed in 341-340 BCE and 60 CE, and earthquakes and tsunamis in 373-372 BCE. Aristotle believed that comets were an indicator of climate and when several appeared within a single year it would be arid and stormy.[22]

During the Roman era, philosopher/statesman Seneca the Younger (4 BCE-65 CE) suggested comets were a kind of star whose movements and location were impossible to predict, and which simply appeared out of the blue.

While arguments raged about what comets were and what their appearance symbolised, there was consensus that they were harbingers of natural disasters such as meteor strikes, droughts and severe winds. The poet/astrologer Manilius described them as "tokens of impending doom".[23]

As well as natural disasters, comets came to symbolise human-related ones, such as riots, wars, and especially the death of a king or a noble figure. It is said that the family of Aeneas, the son of the goddess Aphrodite, witnessed a meteorite in the sky before the Trojan War and treated it as a warning to leave the city. They left and, as a result, escaped the war that destroyed their homeland.

Comets played a special role in Rome in 44 BCE when Julius Caesar was assassinated.[24] Comets were already connected to the idea of Caesar's divinity, but when one appeared months after the emperor's death, his adopted son Octavian, who succeeded him as the Emperor Augustus, announced that the soul of Caesar had ridden on a comet towards heaven. Since Augustus would not have been able to take the throne without Caesar's death, he saw comets as his lucky stars, and enhanced the practice of putting the symbol of a comet on his gold coinage.[25] However, whenever a comet appeared in the sky, Augustus became very sensitive and forbade astrologers from making any predictions on the subject.

Connecting comets with the death of an emperor was sometimes used to undermine his authority. During the reign of the tyrannical Roman emperor Nero, his opponents announced it would be his turn to die when a comet appeared in the sky. Seneca the Younger, the driving force behind this prediction, longed for Nero's death. Not only did Nero survive the appearance of a comet, he ordered Seneca to commit suicide for making such a prediction.[26]

The Roman natural philosopher Pliny the Elder (23-79 CE) gave a detailed description of comets in Book Two of his encyclopaedic *Naturalis Historia*. As well as commenting on their current astrological interpretations, he described and gave names to their different colours and shapes.

He never interpreted these differences astrologically, but others did in later years, including 13th century Italian astrologer and mathematician Guido Bonatti, who paid them close attention and described in detail their possible meaning. Pliny also explored the rumours that comets were associated with the fate of emperors Julius Caesar, Augustus and Nero, and stated that Rome had the only temple in the world dedicated to comets. He also expertly described the two theories current at the time about the origin of comets: stars with their own orbits, or illusions caused by vapour and heat.[27]

The most well know comet is Halley's Comet, named after the astronomer Edmund Halley. When it appeared in the sky in 684 CE, a rainstorm lasted for three months and there was an earthquake followed by an outbreak of the Black Death.

When Halley's Comet passed by again in 1066, William, Duke of Normandy, defeated King Harold II at the Battle of Hastings and claimed the English throne, thus enhancing its association with the death of rulers.[28]

The Danish Renaissance astronomer Tycho Brahe studied the Great Comet of 1577 in greater detail than had ever been done before, making thousands of precise measurements of its path across the sky. His detailed calculations greatly influenced future researchers, most notably Johannes Kepler. The comet was recorded in Peru, Japan, Persia and India and there is a famous engraving of it flying over Prague. Ruler of the Ottoman Empire Sultan Murad III considered the comet a bad omen, blaming it for causing a plague. Meanwhile, according to the Earl of Northampton Henry Howard, England's Queen Elizabeth I caused a stir when, instead of hiding away as she had been advised, she went out to observe the comet and afterwards declared: "Lacta est alia (the dice have been thrown)".[29]

Observing the later comet of 1585, Tycho Brahe made detailed predictions from an astrological perspective. His thoughts, which he recorded in his famous book, paint a picture of how 16th century astrologers imagined the way comets influenced life on earth.[30]

The 1680 comet was the next one to have a major impact on scientists, including astronomers Gottfried Krich, Georg Dörffel, Isaac Newton, John Flamsteed and Edmund Halley. Calculating the comet's orbit, Newton was able to verify Kepler's laws of planetary motion. Later he and Halley used this data to confirm Flamsteed's belief that the same comet appeared the following year in 1681, despite having initially disputed the theory.[31]

Halley's Comet was named after Edmund Halley used Newton's new laws to calculate the gravitational influence of Jupiter and Saturn on comets in general. He claimed the orbit of the comet of 1682 was almost identical to the comets that appeared in 1531 and 1607, and used his calculations to predict its reappearance in 1758. Sure enough, the comet was spotted on 25 December 1758 by German astronomer Johann Georg Palitzch, reaching perihelion on 13 March 1759. That year French astronomer Nicola-Louis de Lacaille named the comet after Halley, though it was too late for the Astronomer Royal to appreciate the honour as he had died 18 years earlier.[32]

Comets' reputation for spelling doom returned in 1910 with the return of Halley's Comet. Panic spread when people were told that the comet's tail contained carbon monoxide and carbon dioxide and its poisonous gases would kill everyone. People blocked up windows and bought pills to protect themselves, but no victims of the comet were ever reported.[33]

Similar rumours were reported by missionaries working in China where people kept their doors and windows closed for fear of their lives. The 1910 appearance of Halley's Comet was also considered to be an omen of the end of imperial rule in China with the fall of the Qing Dynasty in 1912 after a plague and a revolution.

Now that the science behind comets is better understood, we astrologers may be laughed at for still suggesting they are harbingers of disaster. However perhaps we should not ignore those significations attached to past comets, as their appearance still seems to coincide with catastrophe.

In 1882 a Great Comet flew across the sky when a cyclone in the Arabian Sea caused flooding in Bombay harbour, leaving around 100,000 dead.

Two comets seen in January 1910 were followed that summer by race riots in the USA triggered by African-American boxer Jack Johnson's defeat of his white opponent James J. Jeffries in what was dubbed "The Fight of the Century".

A month after a comet passed by in December 1927, the Great Flood of London saw the city inundated with water when the river burst its banks, leading to the death of 14 slum dwellers and leaving thousands homeless. Six months later the small town of Potter, Nebraska, was hit by a record-breaking hailstorm with grapefruit-sized hailstones, the largest measuring a record 17 inches in circumference and weighing 1.5lbs. Two months earlier a similar hailstorm killed livestock in Alabama leaving enough hail on the ground for people to make ice cream with for the next two months.

In 1966 Uranus was conjunct Pluto opposing Saturn and Chiron when Comet Hyakutake passed by. That year saw Chairman Mao Tse Tung launch China's Cultural Revolution by the end of which between 500,000 and 2 million people were said to have been killed and the lives of millions more ruined, thrusting China into 10 years of turmoil, bloodshed, hunger and stagnation.

With such a history, it is not surprising people associate comets with human and natural cataclysms, and fear their appearance in the sky as a portent of disaster.

Chapter 3

Ancient Astrologers on Comets

Famous astrologers of the past such as Ptolemy, Sahl ibn Bishr, Abu Mashar, Guido Bonatti, Tycho Brahe and William Lilly all had something to say about the influence comets have on life on Earth. Tycho Brahe notably made detailed observations of the comets in 1577 and 1586, calculating their orbit and location in the heavens, comparing them in importance to the lunar phases and the Great Conjunction of Jupiter and Saturn every 20 years. Here are some of these astrologers' comments on comets:

Tropical zodiac sign

The element of the sign through which a comet passed was given special attention. Fire signs were associated with fire hazards, conflagrations, wars and slaughter; water signs with extreme rainstorms and tsunamis; earth signs with droughts and food shortages; and air signs with hurricanes, tornados and cyclones.

In 1586 Tycho warned there would be inundations of water and shipwrecks when that year's comet passed through Pisces, and as it moved into Aries he said there could be extreme temperatures.[34]

Traditional astrologers believed every country was ruled by one sign of the zodiac, and when a comet passed through that sign the country could suffer disasters. In 1678 William Lilly suggested that year's comet would impact Poland, Russia, Sicily, Norway, Algeria, Lorraine and Rome as it passed through the sign of Taurus.[35]

Ptolemy always made a point of noting which sign a comet first appeared in.[36]

In the plague year of 1665, English astrologer John Gadbury published his volume *De Cometis* in which he listed how a comet would impact people's lives according to the sign of the zodiac it appeared in.[37] His significations included:

11

Aries: Diseases of the head and eyes, suffering of the poor, sadness and troubles;

Taurus: Diseases and earthquakes, death of great men, damage to livestock, rotten fruits;

Gemini: Children becoming seriously ill, men committing evil deeds, many abortions, strong winds;

Cancer: Famine, plague, war, swarms of locusts or caterpillars, insects destroying fruit;

Leo: Vermin, rats, harm to ladies, many dogs going crazy, corn destroyed by worms;

Virgo: bad for businessmen, noble women prone to scandal and disgrace;

Libra: thieves, robbers, heat, cold, death of kings;

Scorpio: Disputes between men, wars, rebellions, scarcity of grains and fruits;

Sagittarius: Depression for noblemen;

Capricorn: Adultery and fornication prevalent among men, persecution of religious people;

Aquarius: The spread of plague, terrible and long wars, the death of an outstanding prince or a great woman;

Pisces: Many unusual things, destruction of fish.

Fixed Stars and Constellations

Ancient astrologers would also take into consideration the actual constellations and fixed stars where a comet showed up in the heavens. For example, a comet becoming visible in the constellation of Taurus might connect it somehow to cattle; Pisces might relate to the ocean or fisheries; Sagittarius might raise concerns about wars of religion.

In his 1619 book *An Astronomical Description of the Late Comet*, astronomer John Bainbridge confidently predicted the impact of the previous year's comet as it passed through the constellations Ursa Major, Corona Borealis and Serpens.[38]

In his notes on the comet of 1586, Tycho Brahe described the potential influence of it passing through the Beehive Cluster, also known as Praesepe.[39]

Casting a comet's chart

To cast a chart for a comet astrologers had to consider the Ascendant and house positions, which is not easy and was virtually impossible in the pre-technological era. A common practice in traditional astrology was to use 'syzygy' by casting a chart for the New or Full Moon prior to its initial sighting, as Tycho Brahe did in his book where he placed the comet in the 8th house and predicted a disaster that would kill thousands of people.[40]

Tycho also worked with the Moon's waxing and waning squares using the ancient technique 'octatopus' which divides the chart into eight segments, sometimes called the '8 House System'.

In contrast, 17th century English astrologer William Lilly drew up a chart for the first New Moon *following* the appearance of a comet.[41]

Tycho also determined the house a comet should occupy by casting a chart for the moment it crossed the ecliptic. He wrote: "The fact that a comet was also in the eighth house at the same time when it crossed the ecliptic confirms, the majority of Astrologers agree, that the place of Heaven has a deadly and mournful denunciation."[42]

Interpreting a comet's chart

One straight forward technique they employed was noting what planets the comet passed while still visible from Earth. For example, if a comet passed by Mercury it might raise issues related to transport, such as bridges.

In his records of the 1586 comet, Tycho noted the comet aligned with Saturn in the 9th house near the MC. He suggested that would bring powerful storms and chronic diseases.

In his book of 100 astrological aphorisms *Centiloquium*, the 2nd century Alexandrian philosopher and scientist Claudius Ptolemy noted:

"If comets, whose distance is eleven signs behind the Sun, appear in angles, the king of some kingdom, or one of the princes or chief men of a kingdom, will die. If in a succedent house, the affairs of the kingdom's treasury will prosper, but the governor or ruler will be changed. If in a cadent house, there will be diseases and sudden deaths. And if comets be in motion from the west towards the east, a foreign foe will invade the country: if not in motion, the foe will be provincial, or domestic."[43]

Interpreting a comet by its colour

Astrologers in times past would also consider a comet's colour in their interpretation. For example, if it had a grey tone they might suggest its influence would have a Saturnian flavour.

In his famous work *Christian Astrology*, William Lilly made the following connections between a comet's colour and its planetary association:

> Saturn - pale and ashy
> Jupiter – bright, clear, azure
> Mars – shiny, fiery, sparkling, red
> Sun – saffron, gold
> Venus – white, milky
> Mercury – this is not clear
> Moon – white, pale green, silver[44]

Comets were known to change colour sometimes as they crossed the night sky. In 1680 William Knight pointed out that year's comet first appeared with a grey tone, relating it to Saturn and suggesting it would impact the world of politics and resources. After several months it brightened to resemble Jupiter, which coincided with an increase in stresses around religion and law. Finally it turned red, interpreted by Knight as an indication that incidents of theft and robbery would escalate.[45]

A comet's colour depends on its chemical composition, with different chemicals manifesting different shades. For example, calcium will appear with a purple or violet hue, while magnesium turns green or teal. It is not unusual for a comet's tail to appear green. This is due to the gases produced by compounds inside the comet, mainly hydrogen, oxygen, carbon and nitrogen. When cyanide (CN - a carbon/nitrogen bond) or diatomic carbon (C2 - carbon-carbon bond) are exposed to ultraviolet light, their electrons are energised and eventually appear cyan and blue-green.[46]

Modern astrologers have to be careful when interpreting a comet's colour using a photograph. For technical reasons, a camera image will often automatically make celestial objects look green, largely due to the

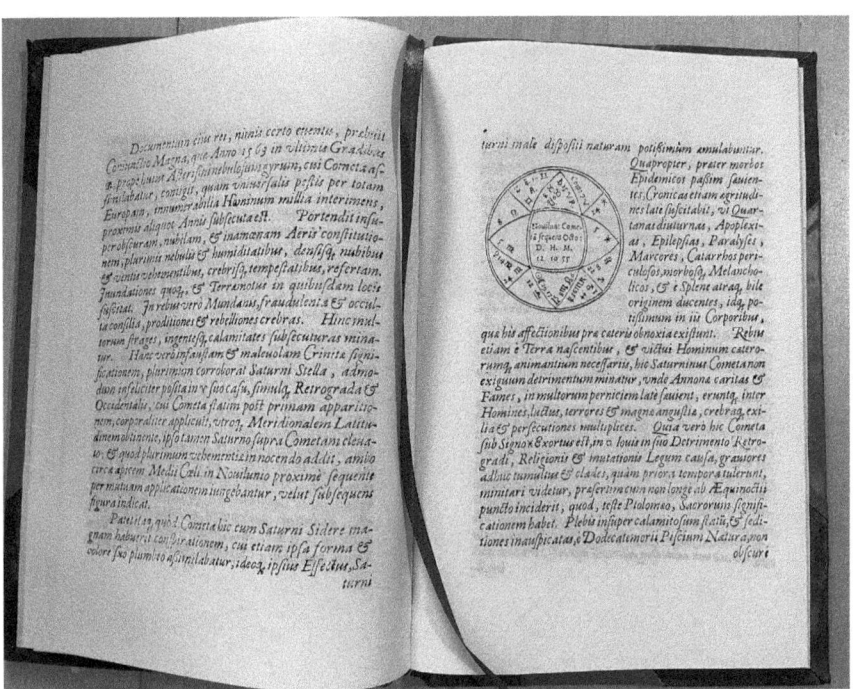

Tycho Brahe Diarium astrologicum et metheorologicum anni a nato Christo 1586 Uranienborg: 1586.

Photo taken by Rod Chang

One Shot Colour camera sensor having one red, one blue and two green pixels.[47] Unless astrophotographers recalibrate the colour of their images, most comets will appear green in colour. Astrologers should base their interpretations on actual sightings and not photographic images.

Direction of the tail

Astrologers of old would also give weight to the direction of the coma, implying it would point towards the location of any imminent disasters. Also any country ruled by the sign towards which the tail pointed could be impacted. To quote Ptolemy:

> "The direction and inclination of their trains, point out the regions or places liable to be affected by the events which they threaten; and the form of the signs indicates the quality and nature of those events, as well as the genus, class, or kind, on which the effect will fall."[48]

16   Comets in Astrology

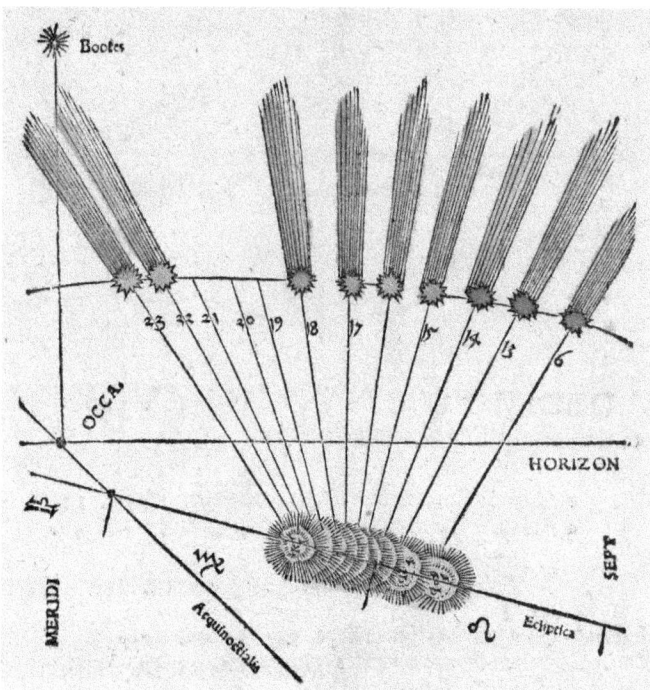

Illustration on page 89v of *Astronomicum Caesareum* (1540). Portrays the tail of a 1531 comet (likely Halley's Comet) always pointing away from the Sun.

(public domain) Attribution: National Library of Poland, Public domain, via Wikimedia Commons

Interpreting a comet by its shape

Comets come in different shapes, leading to a range of different interpretations of their meaning. A fireball with a long coma was described as a "long sword" suggesting war, while a short coma was a "dagger" symbolising assassination. If a comet appeared as bright as the Sun it suggested a great person would be born.

Pliny the Elder described various types of comet, without offering any astrological interpretations.[49] These included:

> Crinitae - shaggy with bloody locks, surrounded with bristles like hair;
>
> Pogoniæ (πωγωνίος) – as if a beard was hanging down from the comet;
>
> Acontiæ (ἀκόντιον, or jaculum) - vibrates like a dart with a very quick motion;

Xiphiæ (ξίφος, ensis) - pale and shiny like a sword and without any ray;

Discei (δίσκος, Orbis) - amber in colour with a few rays emitting from its edges;

Pitheus (πίθος, dolium) - a form of cask, convex and emitting a smoky light;

Cerastias (κέρας, cornu) – shaped like a horn;

Lampadias (λαμπάς) - like a burning torch;

Hippias (ἵππος, equus) - like a horse's mane, fast moving like a revolving circle;

White comet with silver hair, very brilliant and appearing like a deity in human form;

Some comets appear shaggy like a fleece with a crown, while some have the appearance of a mane and change shape into that of a spear.

The 13th century mathematician, astronomer and astrologer Guido Bonatti gave nine descriptions of comets that did not fully accord with Pliny's account. Unlike Pliny, Bonatti also included astrological interpretations (compiled below). For a detailed account of the ancient astrologers' interpretations of comets, see Benjamin N. Dykes, PhD, *Astrology of the World, Volume I*.[50]

Javelin, also known as Veru - a comet appearing close to the Sun and during the day symbolising the death of the king and a decrease in crop yields;

Coenaculum - the colour of Mars, followed by a light that looks like smoke under ashes, symbolising scarcity, but not famine, also a symbol of religious wars;

The Measuring Rod - looks like a rod being dragged along, the light thick, but not very bright, signifying drought, water shortages and food shortages;

Soldier - Venusian in nature and large in size, symbolising people's expectation of change, notably to laws and ancient traditions, as well as harm to kings and nobles;

Ascona - a blue, Mercury-like nature, this comet is small with a long tail and rays extending like wings, symbolising battles and the death of kings and nobles;

Morning/Dawn - red with a long tail and the nature of Mars, symbolising war, fire and famine;

Silver - related to Jupiter and if connected to Jupiter's sign and if Jupiter is dignified, symbolising a good harvest;

Rosy - large and round, coloured somewhere between gold and silver with the appearance of human faces in it, symbolising the death of kings and dignitaries and the emergence of great people, or bringing about general improvements;

Black - Saturn's colour, symbolising death by natural causes or death by beheading.

The 17th century French astrologer Jacques Gaffarel described comets being shaped like hieroglyphs that could be interpreted accordingly. In his 1629 book *Curiositez inouyes sur la sculpture talismanique des Persans, horoscope des Patriarches et lecture des estoiles* (Unheard of curiosities concerning the talismanical sculpture of the Persians, the horoscope of the Patriarchs, and the reading of the stars), he wrote:

"Comets be the Handiwork of God or whether they be but mere Exhalations or lastly, suppose them to be New Stars, appearing in the Heavens; we shall in the next place proceed to shew, that their Figures are Mystical Characters or, as it were, a kind of Hieroglyphics, by which we may be able read, by vertue of Analogy, what Good, or Evil accidents shall befall us. The General Rules whereof do here follow.

"The First is, that if they are figured like a Column, or Pillar they denote the Constancy of some Prince, or of some Great Saint.

"The Second Rule is, that when a Comet, Or fiery Meteor, is Round, Cleare, Bright, and not Duskie at all, but lookes as it were, like another Sun; it may Signifie the Birth of some Great Prince.

"The Third is, that if they be of a Pyramidal Figure, we shall then suffer great Losses by Fire; and, by way of Analogy, may conclude, of some Tyranny approaching. And this is the Opinion of Cornelius Gemma, who expresseth himself in these words.

"The Fourth is; that if they be of much Extent, Waved, and Dissipated here and there in the form of waters; they then denote seditions in the people.

"The Fifth is that if they be of the figure of a Horne, which is the hieroglyph of power; as many be observed out of the Scriptures, in very many places: they foreshew the Great re strength of some Monarch, and an Absolute Power.

"The Sixth is that if they bear the figure of a sword they presage Desolations, which shall be caused by the Sword."

"If the Comet be figured like a Trumpet; it then also foretells of Wars: but if it be of the form either of a Dart; or Arrow; or else of a Javelin, it denounceth both Warre, and Pestilence; the Effects whereof flye abroad; as swift as an Arrow. Such a one; as this, appeared in 80".[51]

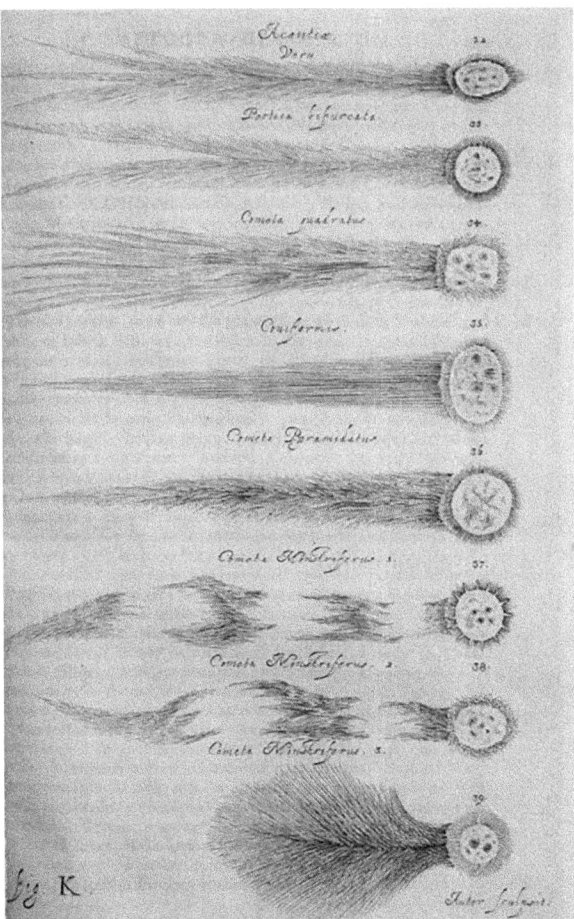

Plate K from the Cometographia by Johannes Hevelius

Johannes Hevelius, Public domain, via Wikimedia Commons

Chapter 4

Mythology

Greek and Roman Mythology

In Greek mythology there are only three stories I have found that relate to comets.

In the first we find the origin of the word 'meteor'. Menippe and Metioche were the daughters of Orion the hunter who were raised by their mother after he was accidentally killed, according to some versions of the myth by the goddess Artemis. It was said the two girls were blessed by the goddesses - Athena taught them the art of weaving, and Aphrodite gave them beauty. When a plague broke out in their homeland of Aonia, the oracle of Apollo announced it would only end when two maidens were sacrificed to appease two of the three Erinyes, or Furies, the angry underworld goddesses of vengeance. When the two sisters heard of the oracle and that no maiden had offered herself up for sacrifice, they put themselves forward to relieve the suffering of their fellow countrymen and women, and stabbed themselves with their weaving shuttles. After their deaths, Hades and Persephone, the king and queen of the underworld, took pity on them and turned them into comets.[52]

A separate myth tells the story of Electra, one of the seven sisters of the Pleiades. She bore a son to the god Zeus named Dardanus, who built up the empire of Troy. Electra brought to the city a wooden statue of the goddess Athena, the Palladium, as a talisman to protect it from harm, but during the Trojan War the Greek warriors Odysseus and Diomedes managed to steal the statue by entering Troy through a secret passage, thus leaving the city unprotected and open to the deceit of the Trojan Horse. According to some accounts, in her grief at the destruction of the city and the death of her son's descendants, Electra tore out her hair and became a comet. Thus Electra is deemed by some authorities to be the one of the Seven Sisters of the Pleiades that cannot be seen, while other myths suggest it to be Merope.[53]

Finally we have a shooting star appearing in the myth of the destruction of Troy in the epic poem of 'The Aeneid' told by 1st century BCE author Virgil.[54] In the story, after the Greeks have outfoxed the Trojans and entered the city by hiding inside the Trojan Horse, the hero Aeneas wakes from a dream telling him to leave the city with his family. He witnesses the destruction taking place, initially joins the fray, and realising the hopelessness of the situation, he gathers his son, wife and father and flees while above him thunder resounds and a 'comet' flies across the sky.

There are different myths that describe the appearance of comets when Troy was captured, which makes me suspect whether comets really did appear at the time. We simply do not have enough data to confirm these things, but these myths have certainly helped me understand how to interpret comets astrologically.

Chapter 5

A Modern Perspective on Comets

So far we have learned how ancient astrologers interpreted the meaning behind the appearance of comets by observing their colour and shape, the constellations and zodiac signs they passed through, the planets and fixed stars they aligned with and by drawing up charts for them. We have also seen that comets were mostly associated with death and disaster. While it is undeniable that both human-related and natural disasters have often occurred when a comet becomes visible, I would like to present a new perspective on these messengers of the gods based on concrete data and drawing on both psychological astrology and Greek mythology.

Symbolism

We can use symbolism based on our modern understanding of the science surrounding comets to find new ways of interpreting them astrologically.

Firstly, where do comets come from? Like the asteroid/planetoid/comet Chiron, they come from the distant outskirts of the solar system beyond the orbits of Neptune and Pluto. This dark, freezing boundary of the solar system has been compared to the collective unconscious by many psychological astrologers, notably Melanie Reinhart.[55]

Secondly, what are comets made of? Symbolically, they could be said to combine all the four traditional elements, being formed from dust (earth), ice (water) and gases such as carbon monoxide and carbon dioxide (air) and then being heated up by the Sun (fire), thus releasing the frozen gases stored inside them.

And thirdly, when our solar system was in its infancy there were comets and meteorites flying about everywhere, colliding with planets left, right and centre. These celestial missiles brought with them water, complex molecules and the building blocks of DNA, and are therefore quite likely to be the source of life on planet Earth.⁵⁶ Thus comets can be associated with evolutionary breakthroughs.⁵⁷

From these scientific facts we can build a modern symbolism to help us interpret the astrological meaning of comets.

Collective fear

The fear and trepidation surrounding the appearance of comets results from the fact that they coincide with major attention-grabbing events that tend to disturb the public consciousness. Psychologically this could symbolise issues arising from the depths of the collective unconscious. As the comet closes in on the Sun (consciousness), the frozen elements within it are heated up and released, thus issues that have been frozen in the collective psyche are suddenly brought to light in dramatic fashion. For example, a mass mortality event or the death of a major public figure might bring to the surface fears of our own mortality.

Comets appeared at the time of the death of major figures including Julius Caesar, Alexander the Great, George V of England, Venezuelan President Hugo Chavez and French President Francois Mitterand. Comets were visible when Charles and Diana divorced in 1996 and when she died in a car crash the following year, and also when both Dutch Queen Beatrix and Pope Benedict XVI abdicated. A comet appeared the year before Britain's longest reigning monarch Queen Elizabeth II died in 2022.

Collective shifts

As well as frozen emotions (water) being released when comets appear, radical new ideas and attitudes (air) also enter into the zeitgeist.

For example, in August 1882, shortly before that year's comet showed up, the Married Women's Property Act came into force in Britain, which, remarkable though it may seem today, allowed women to buy, own and sell property and keep their earnings for the first time.

At the same time France enacted legislation that introduced compulsory education for all children aged six to thirteen, free of charge and free of religion.

In 1910, the year people feared they would be killed by the toxic gases emanating from that year's comet, China made slavery illegal.

Comets were around when the Treaty of Rome was signed in 1957 to establish the European Economic Community (EEC), and also when the EEC signed the Schengen Agreement in 1985 that guaranteed free movement between member countries for the first time.

In 1965 when Comet Ikeya-Seki appeared, the British House of Lords proposed the decriminalisation of homosexuality in the UK.

In 2013, the year of Comet Pan-STARRS, the US, Brazil, France, Uruguay, Britain and Luxembourg allowed same sex marriage for the first time, while Pope Francis said of gay members of the Catholic community: "If they accept the Lord and have goodwill, who am I to judge them?" As a result many gay worshippers felt a sense of guilt or sin being lifted.

Breakthroughs and innovation

Comets often coincide with important scientific, technological and political breakthroughs. They seem to play a critical role with human innovation of all kinds.

Are we really to believe a falling apple awakened Isaac Newton to the idea of gravity while he was studying the gravitational pull of the planets on comets' trajectory through space? Newton possessed a huge collection of books documenting the ancients' studies of these celestial phenomena and had proposed his own model of the impact of gravity in the solar system. It was while observing and calculating the movement of the Great Comet of 1680 that he had a flash of insight that the mass of the Sun played a role in the comet's movements. It was through these studies that he learned to understand the magnetic forces at work in the solar system, work which later helped astronomer Edmond Halley predict the orbit of the comet that eventually bore his name.

One of the most influential events in modern history took place as the Great Comet of September 1882 passed by, when Thomas Edison switched on the first ever electrical generator in New York, an event which led to a transformation in modern lifestyles. Interestingly, the comet appeared in the sign of Virgo, representing the activities of daily life and industry.

The appearance of Halley's Comet in Aquarius and a second Great Comet in Gemini in 1910 coincided with several breakthroughs in their respective fields.

Infrared light had first been discovered in 1800 by William Herschel, who is more widely known, especially in the astrological community, as the man who discovered the planet Uranus with the help of his sister Caroline, 19 years earlier.

Infrared photography only came into being in 1910 when physicist Robert Williams Wood used experimental film requiring incredibly long exposures to produce the first photographs of their kind. It became widely used for aerial photography by the US military during World War I to spot enemy positions through atmospheric haze.

A number of world firsts in aviation took place that year, including the first bombing experiment using sand bags dropped on a target in Los Angeles; the first flight of a seaplane in France; the first flight to climb more than one mile took off in New Jersey; a gun was fired from a plane for the first time near New York City and the first air cargo mission was carried out when Philip O. Parmalee delivered two bolts of silk from Dayton to Columbus, Ohio, on a Wright B-10 bi-plane.

1910 also saw the establishment of Florists' Telegraph Delivery, a ground-breaking service set up by 13 florists serving different cities in the USA who agreed to deliver flowers for each other's out of town customers using the telegraph communication system. With the Great Comet appearing in Mercury's sign of Gemini, it is notable that the company, which has since grown into an international service as Florists' Transworld Delivery, uses the famous 16th century image of the Greek god Hermes, designed by Flemish sculptor Giambologna, as their company logo. Except in their case, Mercury Man is holding a bunch of flowers as he runs forwards.[58]

When Halley's Comet visited again in 1986, the first computer virus appeared.[59] It was called Brain and was devised by two brothers Basit and Amjad Farooq Alvi who ran a computer store in Pakistan. Tired of customers making illegal copies of their software, they developed the virus to infect 5.25" floppy disks.

That year a consortium of British and French corporations and banks announced plans to build a tunnel under the English Channel connecting the two nations, with work starting a year later. When it opened in 1994, the Channel Tunnel was the longest undersea tunnel of its day measuring 50 kilometres.

In 1962 when Comet Seki-Lines was passing by, the first television images were broadcast live by satellite between the United States and Europe using Telstar 1, launched by NASA at Cape Canaveral in July of that year. The first satellite broadcast featured US TV presenters Walter Cronkite and Chet Huntley in New York, and the BBC's Richard Dimbleby in Brussels, as well as images of the Statue of Liberty and the Eiffel Tower and a segment of a baseball game. The power and influence of the technology became apparent when President John F. Kennedy made an appearance in which he denied rumours that the US would devalue the dollar, which immediately strengthened the currency on world markets. Subsequently Cronkite remarked: "We all glimpsed something of the true power of the instrument we had wrought."

It was also the year when the world woke up to the threat of the possibility of nuclear conflict between the USA and the Soviet Union during the Cuban missile crisis. Global annihilation was narrowly averted by the diplomatic efforts of presidents Kennedy and Nikita Khrushchev.

In 1976 when Comet West passed by, Steve Jobs and Steve Wozniak sold their first Apple I personal computer kit.[60] In 1986 when Halley's Comet dimly appeared, Jobs left Apple Inc. and bought The Graphics Group, later renaming it Pixar.[61] In 1996 when Halley's Comet appeared again, he returned to Apple Inc. and introduced the iMac. And in 2007 when Comet McNaught visited, Apple Inc. introduced the first iPhone.[62]

The appearance of Comet Leonard in 2021 coincided with major breakthroughs in artificial intelligence (AI) in a wide range of sectors from computer games to military technology, with major companies like Facebook, Amazon and Google all announcing new developments. This was the year when the public really woke up to the potential impact that AI could have on society in the future.

A different type of innovation took place in the financial sector in the unlikely setting of Cyprus. In 2013 Comet Pan-STARRS turned up when a financial crisis overwhelmed the Mediterranean island as part of the domino effect of the global financial crisis of 2008. After much debate and hand-wringing, an innovative €10 billion bailout was agreed, jointly funded by the European Commission (EC), the European Central Bank (ECB) and the International Monetary Fund (IMF). In return the Cypriot government agreed to close the island's second largest bank, impose a levy on all uninsured deposits and seize almost half of those deposits in the island's largest commercial bank, much of which belonged to Russian oligarchs who were using Cyprus as a tax haven. There was a panic amongst regular citizens when they found themselves unable to withdraw any money for a whole week after banks were closed to avoid a bank run.[63]

The appearance of comets also correlates with significant scientific discoveries in the realm of biology and medicine.

Physician and microbiologist Robert Koch, nicknamed the father of microbiology, was awarded the Nobel Prize for his discovery of the tuberculosis bacterium on 24 March 1882, when that year's Great Comet passed by. The date is now celebrated every year as World Tuberculosis Day. Koch also discovered the cause of anthrax and cholera and changed the course of modern medicine by introducing germ theory, thus creating the scientific basis of public health.

When German physician Alois Alzheimer first presented his findings about the form of pre-senile dementia that was named after him in the year 1906, his work was not officially recognised. The disease was not named until his ideas were published by his colleague Emil Kraepelin in 1910, the year of the two Great Comets. By the following year the term Alzheimer's Disease had entered the public consciousness and was being used widely in the medical community.

When Comet Skjellerup-Maristany appeared in the sky over the Christmas period in December 1927, receiving a mention in JRR Tolkien's book *Letters from Father Christmas*, two of the earliest breakthroughs were made in genetics that were only confirmed by scientists in later years. That year Russian biologist Nikolai Koltsov proposed that a 'giant hereditary molecule' made up of 'two mirror strands' was responsible for inherited traits in people, an idea confirmed by the discoverers of the double helix structure of DNA, Crick and Watson, in 1953.

Meanwhile in January 1928, the month the comet disappeared from view, British bacteriologist Frederick Griffith reported what became known as the 'Griffiths Experiment', in which he demonstrated the phenomenon of 'bacterial transformation', one of the first examples of the central role DNA plays in heredity, thus indirectly proving its existence.[64]

Unpredictability

The unpredictability and element of surprise that comets display have added to their astrological interpretation since ancient times. Even today, skywatchers are unable to accurately predict a comet's exact path or level of visibility. A classic example of this came in 1986 when Halley's Comet reappeared in the sky after 76 years. Its previous display in 1910 had been spectacular, so when astronomers first detected its approach in 1982 there was considerable media excitement about the light show people could expect. However when it did finally appear four years later, the comet was on the opposite side of the Sun to the Earth, and thus provided what was probably its worst ever light show of the past 2,000 years.

Similar disappointment followed the appearance of Comet Kohoutek in 1973 after it had been dubbed the "Comet of the Century" by an overexcited press. The comet was first discovered by Czech astronomer Luboš Kohoutek further away from the Sun than any previous comet, leading to great anticipation among scientists and the general public alike that it would be the brightest anyone had seen in their lifetime. However in the end it was barely visible to the naked eye, and even then for only a very short period in early 1974.

Mythical correlations

In the myth of the sacrifice of Menippe and Metioche, we see not only death, but also the shape of a weaving shuttle resembling a comet. The main theme of this story is sacrifice.[65] It is a theme that repeatedly manifests on the world stage when comets appear. For example, Diana, Princess of Wales, always carried an aura of sacrifice. The month she was born in July 1961, a South African Airways flight attendant observed Comet Wilson Hubbard. The Great Comet Hale-Bopp was visible to the naked eye for 18 months between 1996 and 1997, the period that included both Diana's divorce from Prince Charles and her death in a car crash in Paris. Her death resulted in the entire country being consumed by grief and stirred resentment at the lack of emotion shown by the royal family.

The theme of sacrifice and grief reared its head again in 2013 when Comet Pan-STARRS was spotted by astronomers, even though it never became visible to the naked eye. That year three young women hit the world's headlines when they escaped from the house of their kidnapper and abuser, school bus driver Ariel Castro, in Cleveland, Ohio, after being locked up in a nightmare of hell for more than 10 years. Their story stirred the hearts of millions and raised awareness about the fate of so many missing children around the world.

The theme of grief is also visible in the Greek myth of the goddess Electra and her feelings after the sack of Troy, the centre of her son's ancient kingdom.

An example of the connection of comets with the birth of a new culture can be found if we return to the myth of Aphrodite's son Aeneas, who escaped the destruction of Troy with his family after witnessing a comet. He went on to settle in what is now Italy and became known as the father or grandfather of Rome as an ancestor of the city's founders Romulus and Remus. Thus he symbolised the shift from Greek to Roman culture that came to dominate the ancient world.[66]

Similarly the Norman Conquest of 1066 that coincided with the appearance of Halley's Comet, as depicted in the famous Bayeux tapestry, transformed the nature of Britain from a series of small kingdoms to a single nation ruled by the first in a long line of kings and queens that continues to the present day in King Charles III.

Chapter 6

Comets and Natal Charts

I remember an astrologer telling me that anyone born within a year of a comet appearing in the sky would be more affected than other people when another comet appeared. They suggested the native could be an innovator, unpredictable, highly influential or connected to shocking events in some way, or they may be involved symbolically with the release of significant information, even of people, or may themselves be released in some way.

I am not entirely convinced this is an accurate interpretation, however I do believe that comets signify something when they impact an individual's natal chart. Here are some examples of "coincidences" around comets that I have noticed in my research.

The American author Mark Twain was born in 1835 when Halley's Comet appeared in the sky. The same comet was seen during the year of his death in 1910.

The dictator Ferdinand Marcos became president of the Philippines in late 1965 when the bright Comet Ikeya-Seki was first discovered, and he was forced to resign and flee the country in 1986 when Halley's Comet made one of its regular return trips.

An even more famous 20th century dictator, Adolf Hitler, was born two months before Comet Barnard 2 was discovered in June 1889. The previous year people had seen two comets cross the heavens: Comet Sawerthal in April and Comet e 1888 (Barnard) in December.

Seven years earlier two comets appeared, including the Great Comet of 1882 that was so bright it was visible in broad daylight. That year saw the birth of a surprising number of men who went on to become influential and controversial leadership figures, including:

- The longest serving US president Franklin D Roosevelt, who led the country through the Great Depression and into World War II;

- Iran's controversial prime minister Mohammed Mossadegh, who was murdered after angering western powers when he nationalised the country's oil industry;
- Irish republican Seán T. O'Kelly who became the country's second president after it attained independence from Britain;
- Germany's chancellor Kurt von Schleicher, who was murdered by Hitler's followers during the Night of the Long Knives;
- Swedish King Gustaf VI Adolf who removed the monarchy's last remaining political powers before his death;
- Brazilian dictator Getúlio Vargas, considered to be Brazils' most influential politician of the 20th century despite being surrounded by controversy;
- Thailand's longest reigning monarch who had the third longest reign in the world, Bhumibol Adulyadej, was born in December 1927 when the very bright Comet Skjellerup–Maristany was spotted in the sky.

Comet Arend-Roland was discovered in November 1956 at the peak of the Suez Crisis in Egypt, which saw the humiliation of Britain and became a symbol of renewed confidence in the Islamic world thanks to the leadership of Egyptian president General Gamal Abdel Nasser. Shortly before the comet reached perihelion in April 1957, another influential Muslim leader of a very different stripe was born. Osama bin Laden also challenged the western world through the radical militant group Al Qaeda, claiming responsibility for the 9/11 attacks on the Twin Towers in New York, which rank highly amongst the most shocking events the world has ever seen. And to cap it all, bin Laden was killed during a military raid on his home in Pakistan in 2011, the year of Comet Lovejoy.

Five months after bin Laden's birth, another rebel was born. Chinese artist Ai Weiwei has become famous for his open criticism of the Chinese government's standing on human rights and democracy, and was arrested and held without charge for almost three months the same year that bin Laden died.

Comets in Transit

My research also suggests that comets can have a dramatic impact on our lives if their appearance in some way transits important placements in our natal chart. This can manifest in shocking or sudden incidents occurring, or unexpected shifts in outlook or attitude.

I love the idea expressed by French astrologer André Barbault in his book *L'Astrologie Mondiale* that comets are messengers bringing information and insights from the outer edges of the solar system – the Oort Cloud and the Kuiper Belt – before passing through each planetary orbit on their journey towards the Sun. Their orbits are also irregular, they may be retrograde, they tend not to move along the ecliptic, some comets appear as far north or south as the celestial poles. As such they represent what is irregular in our world, the 'out of the ordinary' in society, the black sheep of the family, so to speak. They can shock and upset the harmony of our universe, or bring ideas that are completely 'out of the box'.

One way of looking at the impact a comet might have on our own natal chart is to simply view it as we would any transiting planet. In my research I have found that certain moments in a comet's journey are more worthy of attention when looking at the way they impact our charts by aspect. These include:

1. The moment the comet was first discovered
2. The time the comet became visible to the naked eye
3. The time it crossed the ecliptic or celestial equator while visible
4. The moment of its perihelion
5. Any alignment with a planet during visibility
6. The time the comet disappeared from view.

Steve Jobs

Let me use Apple founder Steve Jobs as an example of how this might work. Jobs was born at 7:15pm on 24 February, 1955, in San Francisco, USA. Six months earlier, bright Comet 12P/Pons-Brooks was visible, suggesting he would display a certain sensitivity to future comet appearances.

In 1976 Comet West, which has a 500,000 year orbit, was bright enough to be seen in the sky as it passed through the zodiac signs of Capricorn,

Comets and Natal Charts 33

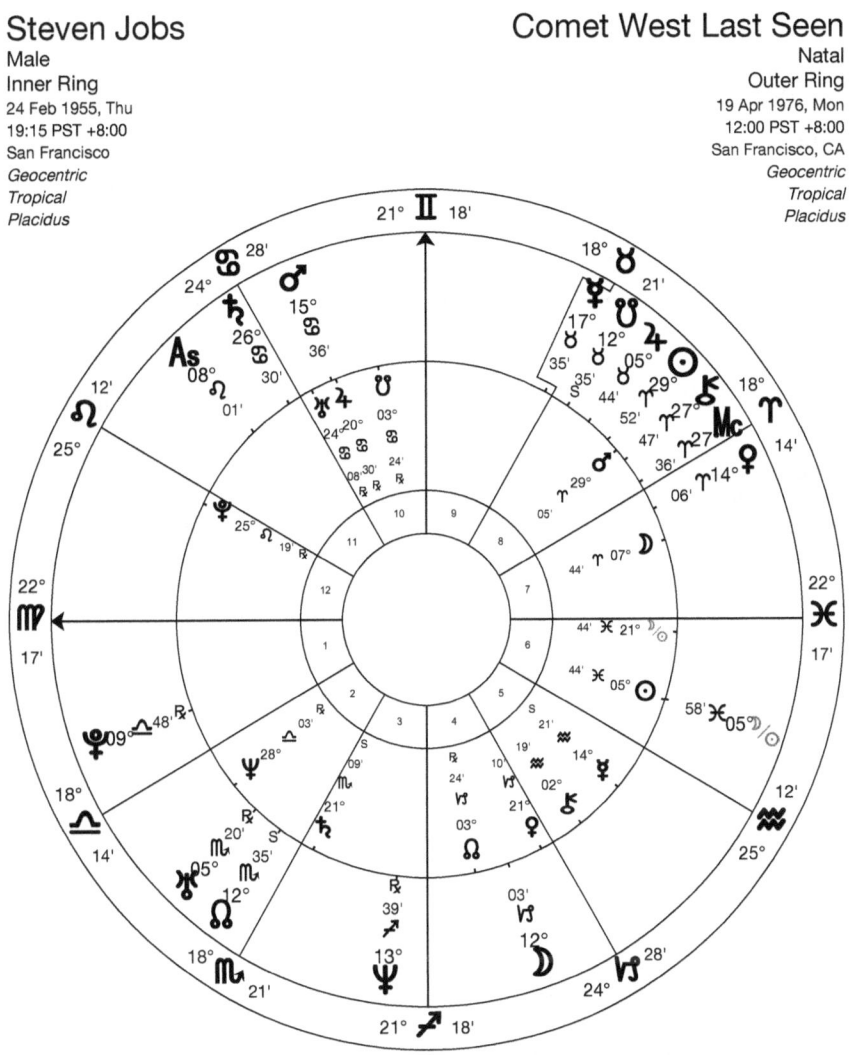

Figure 6.1

Aquarius and Pisces. This journey saw the comet track past Jobs' natal Venus, Chiron, Mercury and Sun and coincided with the period when he and his partner Steve Wozniak were building the first Apple computer. The comet actually disappeared from view when it was at 14 Aquarius, partile with Jobs' natal Mercury, which rules his MC. (Figure 6.1).

In January and February 1986, Halley's Comet became visible once again with its perihelion at 15 Aquarius, again very close to Jobs' natal Mercury. It then aligned with transiting Mercury and Venus at 17 Aquarius, further triggering that critical 14 Aquarius degree for Jobs' planet of communication. (Figure 6.2).

This proved to be a critical period in the innovator's life during which he was forced to leave the company he had founded, leading him to create two new companies that had significant influence on the future of technology. He set up NeXt Computer after his forced resignation in September 1985, which was later used by English computer scientist Tim Berners-Lee to create the World Wide Web at CERN Switzerland in 1990.

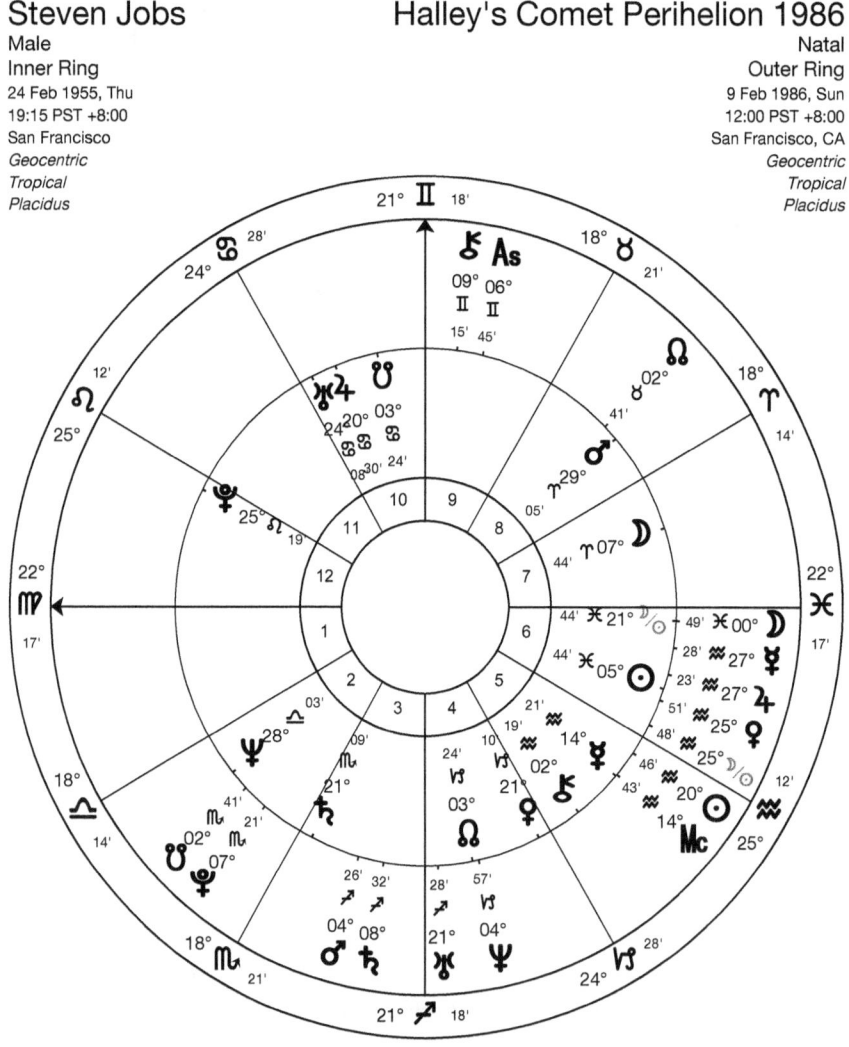

Figure 6.2

Also in 1986 Jobs took over the computer graphics division of Lucasfilm that became Pixar, which produced the world's first 3D computer graphics cartoon *Toy Story*, followed by a series of Oscar winning animations, marking the start of a revolution in the industry.

A further comet connection came with the launch of Apple's iPhone in January 2007 when Comet McNaught, also known as the Great Comet of 2007, was visible in the sky aligning with Jobs' natal Venus at 21 Capricorn. The iPhone stands as probably the most significant revolution in 21st century communications technology so far.

Jobs' life demonstrates how anyone can apply the appearance of a comet to their chart, just as they would a transiting planet, keeping an eye on the above mentioned critical points in the comet's journey across the sky.

Julian Assange

Comets appear to have featured large in the life of another world famous character associated with the world of computer technology, the Australian founder of Wikileaks.

Julian Assange was born in Townsville, Queensland, on 3 July 1971. (Figure 6.3) The previous year there were an unusual number of comet sightings, including three that were visible to the naked eye:

- Comet Bennett (C/1969 Y1), a Great Comet, was the brightest and could be seen from February until May 1970;
- Comet White-Ortiz-Bolelli (C/1970 K1) was visible briefly from 16 May until 1 June 1970. It is what is known as a Kreutz sungrazer, part of a family of comets that broke away from a parent comet several centuries ago; and
- Comet Abe (1970G) was visible for a while in September that year.

My speculations suggest that people born during or around this period would find themselves more influenced than most when future comets appeared in the sky, as was the case with Assange.

The future hacker did not own his first computer, a Commodore 64, until 1987, however he gained access to one in 1982 aged just 11 years via a shop in Goolmangar, New South Wales, where his family had moved. That

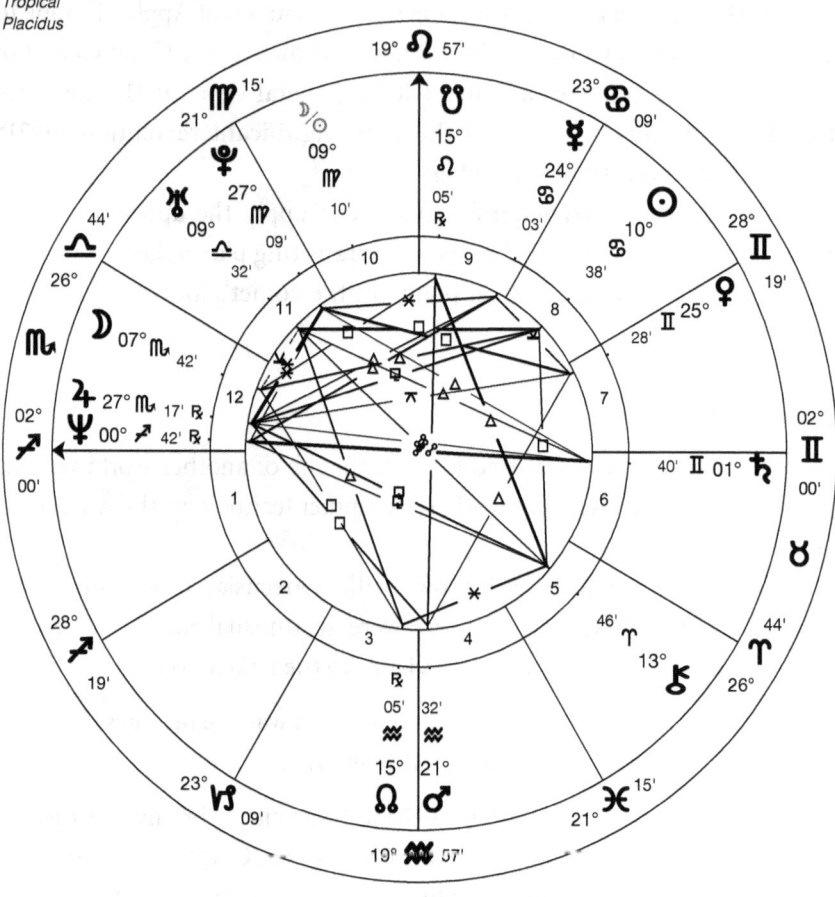

Figure 6.3

was the year astronomer Rod Austin discovered Comet Austin (1982G) on 18 June, becoming visible to everyone during July and August.

It was 1987 when Assange was said to have turned into a skilled hacker, when he was accused of using his computer to steal money from Citibank, though he was never actually charged with theft. He later claimed that aged 17 he had managed to break into the servers at the Pentagon, the headquarters of the US defence department. At the time Pluto was transiting across the teenager's natal Moon at 7 Scorpio, Uranus was opposing his natal Venus at 25 Gemini and Neptune was about to oppose his natal Sun at 10 Cancer, ruling his MC, and perhaps hinting at a future career.

That was also around the time Halley's Comet made its most recent pass close to the Sun, which only happens every three quarters of a century. It reached perihelion in 1986 at 15 Aquarius, right on Assange's North Node, also indicating his future direction in life. In February that year Venus, Mercury and the Sun all crossed that same degree within days of each other, and on 14 April the comet aligned with Pluto at 6 Scorpio, one degree from Assange's Moon – all suggesting it was having considerable impact on the young man.

The next significant period in the hacker's life coincided with the time when Comet Hale-Bopp (C/1995 O1), the brightest comet of recent decades, appeared in the sky for more than 18 months until April 1997, the longest ever recorded.

During this period Assange admitted himself to a psychiatric hospital suffering from depression while he awaited trial facing 31 hacker-related charges, including defrauding Telecom Australia. The trial was set for May 1995, shortly before Hale-Bopp was discovered in July by astronomers Alan Hale and Thomas Bopp at 29 Cancer, close to his natal Mercury. The comet was first observed by the naked eye in May 1996 at 23 Capricorn, directly opposite his natal Mercury.

Assange was facing a theoretical maximum sentence of 290 years in jail, but in December 1996 he pleaded guilty to 24 hacking charges. The judge took into account his disrupted childhood and absence of malicious or mercenary intent, fining him A$2,100 and releasing him on a A$5,000 good behaviour bond rather than sentencing him to prison. The young man later described the experience as formative, leading him on to found Wikileaks. However he was already involved in a wide range of activities related to online security. He had helped the police in Victoria tackle online child pornography, he took over one of Australia's first public internet service providers, he wrote computer programmes, ran a website on computer security and was earning a substantial income as a consultant for large corporations. It was also during the mid '90s that he started working on information 'leaks'.

The Wikileaks domain name was registered in October 2006, posting its first 'leak' the following December. Interestingly the site was inspired by The Pentagon Papers, secret US defence department documents revealing unreported activities by the US during the Vietnam War. Journalist Daniel Ellsberg published the stories in the New York Times on 13 June 1971, less than three weeks before Assange was born.

A number of significant transits were affecting Assange during 2006: the South Node was crossing his natal Pluto, the planet symbolising secrets, scandals, burials and excavation; Neptune was transiting his North Node in Aquarius and Saturn was crossing his Leo MC. It was also the year of his third Jupiter return, often a time of significant turning points for those with Jupiter-ruled Sagittarius rising.

What is interesting for the purposes of this book, is that the unusual Comet Swan (C/2006) suddenly became visible to the naked eye when it flared dramatically on 24 October 2006. The comet had been first discovered by astronomers Robert Matston and Michael Mattiazzo on 20 June at 16 Leo, one degree from Assange's South Node. The South Node's astrological symbolism is connected to release as the tail of the dragon, while the head of the dragon, the North Node, relates to swallowing or devouring. The comet crossed the celestial equator on 17 July at 20 Leo, right on Assange's MC, and travelled from 6-10 Scorpio when it became visible to the naked eye, crossing the Wikileaks founder's Moon.

In addition, Comet McNaught (C/2006 P1), the Great Comet of 2007, was first seen at 10 Capricorn opposite Assange's natal Sun at the time the first 'leak' was published.

Life started to become difficult for Assange in 2010 when internal dissension within Wikileaks led to several resignations, and he was faced with extradition to Sweden to answer allegations of sexual offences, which he consistently denied. In 2012 he sought refuge and gained political asylum in Ecuador's embassy in London where he was to remain until his eventual arrest in 2019. He was subsequently locked up in the high security Belmarsh prison from where he fought extradition to the US on charges of espionage.

Shortly before his arrest, in December 2018 Comet Wirtanen (46P) passed so close to the Earth that NASA considered staging a "comet hopper" mission where a spacecraft would land on the comet for scientific research purposes. The comet crossed the ecliptic at 1 Gemini, conjunct Assange's natal Saturn and Descendant.

After five years of incarceration in one of Britain's most notorious prisons, Assange agreed a plea bargain deal with the US justice department.

He was finally released on 24 June 2024 and returned to Australia two days later, shortly after Jupiter moved into Gemini crossing his natal Saturn and Descendant, while Neptune opposed his natal Pluto and Venus conjuncted his Cancer Sun.

However a comet was also playing a part in the heavens during this fateful year in Assange's turbulent life. Between March and May 2024, as his release was being negotiated, Comet 12P/Pons-Brooks was visible in the sky to many, crossing the celestial equator at that familiar degree of 2 Gemini, conjunct his natal Saturn and Descendant, perhaps assisting his eventual release and path back to freedom. One might also note that when Pons-Brooks was first discovered back in 1812, it was at 23 Gemini, aligned very closely to his natal Venus, and thus a potential celestial friend.

Agatha Christie

The English mystery novelist Agatha Christie is another character who deserves exploration. Christie remains the most successful fiction writer of all time having sold more than two billion of her 66 detective novels. Her stories continue to be made into films and television series, and her play *The Mousetrap* has been running in London's West End continuously since 1952, making it the longest running show in the history of the stage. And her own life is as interesting as her many plots.

Born into a wealthy family in Torquay, Devon, on 15 September 1890, Christie's natal chart has the Sun at 22 Virgo, the Moon at 6 Libra, the Ascendant at 23 Sagittarius conjunct Mars at 24 Sagittarius, with Uranus conjunct her MC at 24 Libra. One can imagine how appropriate this combination would be for dreaming up twisted plots for murder mysteries, with her Virgo Sun symbolising her delight in detail. (Figure 6.4).

Comet Davidson (C/1889 01) appeared in the sky just over one year before Christie was born, just outside the 12 month time frame I have suggested for a comet's influence on a native's birth. It was discovered on 19 July 1889 at 16 Libra aligned with Uranus, bringing in the archetypes of shock, surprise and innovation that were such a feature of her writing. When Christie was born, Mercury was just about to station retrograde at 15 Libra, within a degree of the comet's discovery point.

When the comet was first observed by the naked eye five days later on 24 July, it had moved to 24 Libra, exactly conjunct Christie's natal Uranus

Agatha Christie
Natal
15 Sep 1890, Mon
14:14 UT +0:00
Torquay, UK
Geocentric
Tropical
Placidus

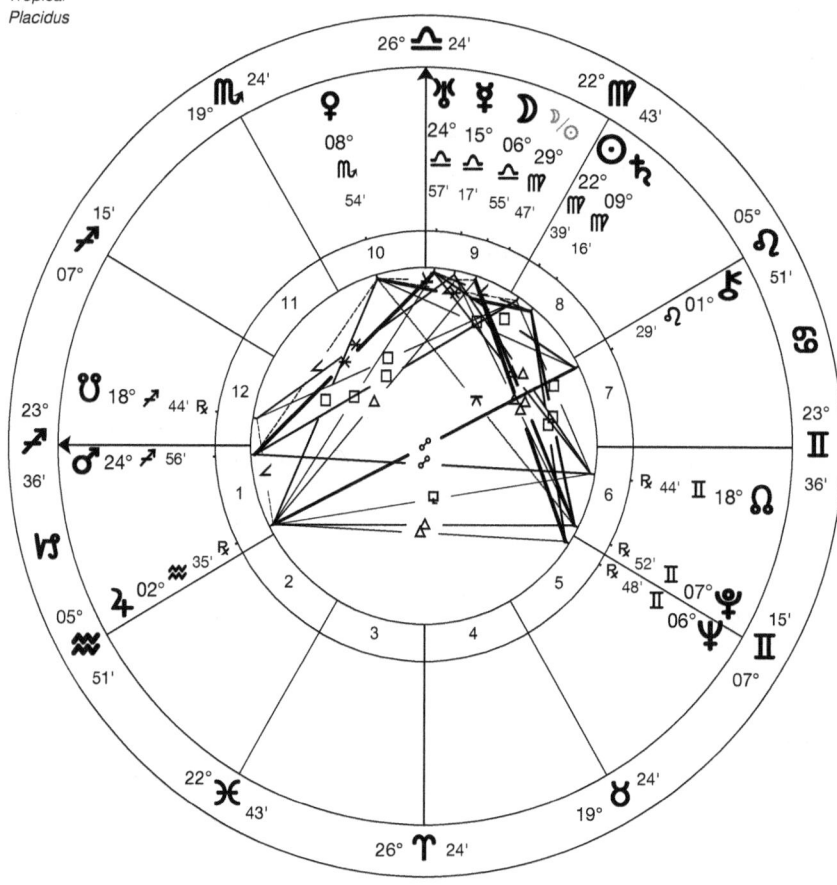

Figure 6.4

and just one degree from her MC. It reached 6 Scorpio by the time it disappeared from view, just two degrees from natal Venus.

Early life

Christie was a voracious reader from an early age and wrote her first poem *The Cow Slip* in April 1901 at the age of 10. Seven months later in November her father Fred died of pneumonia and chronic kidney disease, which Christie later described as being the end of her childhood. That year transiting Uranus conjoined her South Node and sextiled her Mercury, but also the Great Comet of 1901 appeared at 16 Gemini within a degree of transiting Pluto, very close to her North Node and square to her Sun.

During the winter of 1907-8, Christie and her ailing mother spent three months convalescing in the warm climate of Egypt, a popular destination for wealthy English tourists at the time, where she enjoyed many social events. Later many of her mysteries were set in Egypt and the Middle East.

After their return to England while recovering in bed from an illness, 18 year old Christie wrote her first short story *The House of Beauty*, a tale about "madness and dreams", a theme that fascinated her along with so many others born under the rare and powerful Neptune-Pluto conjunction of the late 19th century.

At the time Pluto was crossing her Descendant and opposing natal Mars, while Neptune and Uranus were both squaring natal Mercury from opposite ends of the sky.

Comet Daniel (C/1907 L2) had been visible for three months before the trip to Egypt, aligning with Jupiter in the sky at 1 Leo on 27 August 1907, right on Christie's natal Chiron and opposing natal Jupiter, thus bringing up themes of long distance travel and the inspiration it can arouse.

The following year Comet Morehouse (C/1908 R1) was discovered on 2 September at 8 Gemini, very close to that natal Neptune-Pluto conjunction in Gemini, the sign most associated with writing.

Marriage and World War One

Christie met her first husband Archie Christie in October 1912; they fell in love and were engaged within three months. They married two years later in December 1914 when Archie was home on leave from the front after World War One broke out in August that year. This was a time of great upheaval for Christie who worked as a volunteer nurse at a Red Cross hospital in Torquay, later becoming an apothecary's assistant where she was paid the grand sum of £16/year. Her training supplied her with a rich knowledge of pharmacology that helped in so many of her plots involving poison. Between 1913 and 1914 transiting Uranus was close to natal Jupiter and opposing natal Chiron, while Neptune was triggering the same opposition from the other side of her chart. At the same time Saturn was closing in on her Descendant from the sixth house.

However this was also a period of intense comet activity impacting Christie's natal chart. Shortly before meeting Archie at a dance outside Torquay, Comet Schaumasse (24 P) was discovered on 1 December 1911 at 14 Libra, conjunct her seventh house ruler, Mercury.

Comet Gale (C/1912 R1) was discovered on 9 September 1912 at 6 Scorpio, just two degrees from natal Venus, remaining visible until November shortly before Archie's marriage proposal.

In 1914 as World War One broke out, three bright comets appeared in the sky:

- Comet Zlatinsky (C/1914 J1) visible from 15 May until 28 May;
- Comet Delavan (C/1913 Y1) visible from 16 July until September;
- Comet Campbell (C/1914 S1) visible from 18 September until 10 October.

Comet Zlatinsky had global significance as it aligned with Pluto at 29 Gemini on 23 May and crossed the world axis at 0 Cancer the following day while very close to Earth. It impacted Christie by aligning with transiting Saturn at 18 Gemini on 21 May on her natal North Node and with transiting Venus at 25 Gemini opposite her Mars-Ascendant conjunction.

Comet Delavan appeared in a similar position two months later on the eve of war breaking out. It became visible at 21 Gemini near her natal North Node, opposite her natal Mars-Ascendant and aligned with transiting Saturn at 26 Gemini on 26 July, two days before the war officially started. On 3 August, the day Germany declared war on France and invaded Belgium, the comet aligned with transiting Pluto at 1 Cancer square Christie's natal Moon.

On 5 September the comet aligned with transiting Neptune at 29 Cancer close to Christie's natal Chiron; by 29 September it had reached the South Node at 5 Virgo close to her natal Saturn; and on 12 November it passed retrograde Mercury at 8 Scorpio, the exact degree of the author's MC ruler Venus.

Meanwhile Comet Campbell's brief visit saw it align with the transiting North Node at 5 Pisces on 1 October, opposing Christie's natal Saturn.

During 1917 when she was qualifying as an apothecary's assistant, Jupiter moved from Taurus into Gemini and crossed that natal Neptune-Pluto conjunction, while Pluto squared her Moon and Saturn-Neptune opposed her Jupiter and squared her MC ruler, Venus.

This was the year Comet Mellish (C/1917 F1) was discovered on 19 March and was visible to the naked eye from 11 April at 18 Aries, aligning with transiting Venus at 17 Aries and two days later with Mars at 14 Aries, opposing Christie's natal Mercury at 15 Libra.

Career

In 1920 Christie managed to publish her first detective novel *The Mysterious Affair At Styles*, which she had written four years earlier but had struggled to find a publisher. It first appeared as a weekly serial in *The Times* newspaper between 27 February and 26 June, not coming out in book form until 21 January 1921, and featured Belgian detective Hercule Poirot, one of fiction's most famous characters, who was based on some of the Belgian refugees and patients she met in Torquay during the war.

This was an extraordinary period of planetary alignments for Christie. It was the year of her Saturn return and on the day the first episode appeared in *The Times,* a Sun-Uranus conjunction in Pisces was opposing both natal and transiting Saturn. Natal Venus in Scorpio was conjunct transiting Mars and was being squared by Neptune in Leo. At the same time Christie's Venus and transiting Mars were forming a grand water trine with transiting Pluto in Cancer and the Pisces Sun-Uranus conjunction. (Figure 6.5)

There was also an extra influence from Comet Brorsen-Metcalf (23P), a periodic Halley-type comet with an orbit of 70 years. It was first discovered on 21 August 1919 at 24 Pisces opposite natal Virgo Sun and square her natal Ascendant-Mars conjunction in Sagittarius. Shortly afterwards it aligned with Chiron at 5 Aries opposite her Moon and on 24 September it aligned with retrograde Venus at 13 Virgo, close to her Saturn.

The comet disappeared from view on 21 October 1919 at 27 Virgo, a point just two degrees away from Christie's Sun-Moon midpoint, five degrees from her Virgo Sun and nine degrees from her Libra Moon.

It is of note that Comet Brorsen-Metcalf reappeared in the sky in 1989, the year that ITV began broadcasting their highly successful serialisation of the Hercule Poirot novels, *Agatha Christie's Poirot*, with David Suchet starring as the famous detective. The series ran for 70 episodes over 13 series, finally ending in 2013, but it is still regularly repeated.

44   *Comets in Astrology*

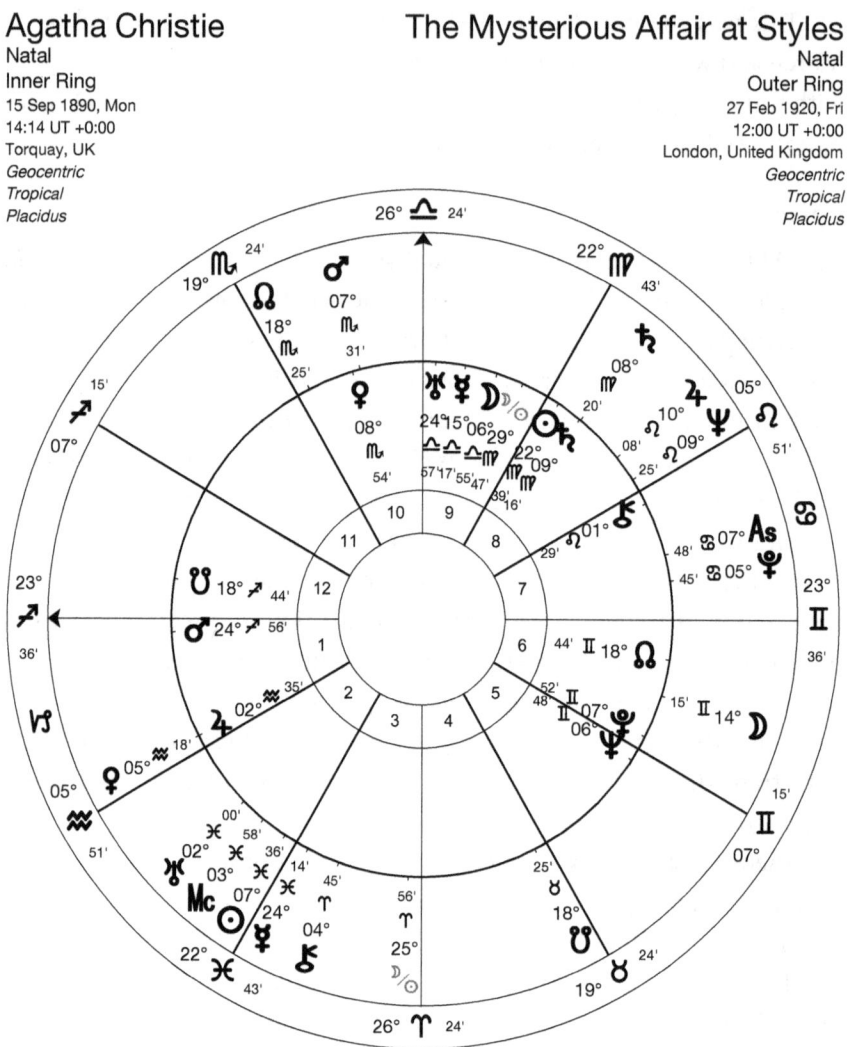

Figure 6.5

Disappearance

The most mysterious and dramatic event in Christie's life occurred in December 1926 when she disappeared for 11 days, sparking an international media sensation.

In April that year her mother, to whom she was extremely close, died, sending her into a deep depression. Things got worse in August when her husband Archie asked for a divorce, declaring he had fallen in love with another woman. The couple quarrelled on 3 December when

Archie announced he intended to spend the weekend away without her. That evening she disappeared and the following morning her abandoned car was discovered above a chalk pit in Surrey and it was feared she had drowned in a nearby pool. The search involved 1,000 police officers, 15,000 volunteers and several aeroplanes scanning the area. One newspaper offered readers a £100 reward (worth a lot of money in those days). The story made the front page of *The New York Times* and Sherlock Holmes' author Arthur Conan Doyle even consulted a medium to find her. She was eventually located at a health spa hotel called *The Old Swan* almost 200 miles away in the Yorkshire town of Harrogate, and claimed she had been in a "fugue state" and had no memory of what had happened.

The significance of the year 1926 in Christie's life is further highlighted as it was the year she published *The Murder of Roger Ackroyd*, described as her masterpiece and voted the best crime novel ever written by the British Crime Writers' Association in 2013.

That year was another powerhouse astrologically for Christie with transiting Uranus opposing her Sun and her Sun-Moon midpoint, transiting Chiron hovering over her Aries IC and Neptune in Leo in sextile to natal Uranus and trine to natal Mars. These planetary configurations reflect the shock, surprise and mystery surrounding both her disappearance and the plot twists in the novel that enthralled so many readers.

Furthermore, as Pluto transited midway through Cancer that year, it squared Christie's seventh house (Placidus) Mercury as she experienced the loss of her husband to another woman, made headlines about her presumed death and (spoiler alert) made Hercule Poirot's assistant both narrator and murderer in her novel– seventh house, Pluto and Mercury themes galore.

Now introducing the comet factor, on 6 May 1925 Comet Orkisz (C/1925 G1) became visible to the naked eye at 2 Aries opposite her natal Moon, aligned with transiting Mercury at 22 Aries on 6 May opposite natal Uranus. The following day transiting Chiron was at 25 Aries close to her natal IC, triggering the foundation of her chart one year before the very foundation of her life was turned upside down with the loss of her beloved mother and her marriage. After the separation she became ill and the couple finally divorced in 1928. Later she wrote: "So after illness came sorrow, despair and heartbreak." (Figure 6.6)

46   Comets in Astrology

Agatha Christie
Natal
Inner Ring
15 Sep 1890, Mon
14:14 UT +0:00
Torquay, UK
Geocentric
Tropical
Placidus

1926 disappearance
Natal
Outer Ring
3 Dec 1926, Fri
12:00 UT +0:00
London, United Kingdom
Geocentric
Tropical
Placidus

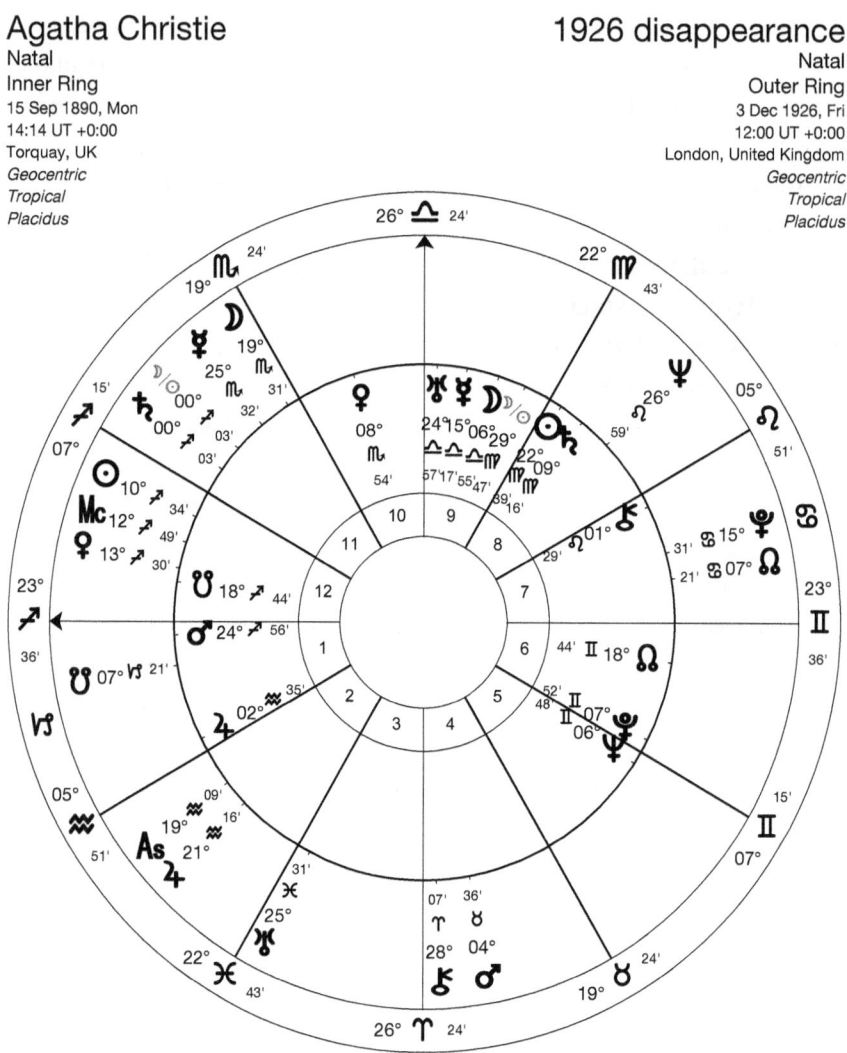

Figure 6.6

Second marriage

Christie met her second husband archaeologist Max Mallowan during a trip to Baghdad, in Iraq, in February 1930. At the time her natal Moon was being opposed by transiting Uranus and squared by transiting Saturn forming a powerful T square, while transiting Jupiter was conjunct her Neptune. Using the astromapping technique, Christie's Jupiter line passes through both her birthplace and Baghdad!

In 1930 there were three comets visible to the naked eye, two of which were discovered by Polish astronomer Antoni Wilk.

Comet Wilk (C1929 Y1) was first observed on 30 December 1929 at 5 Aquarius, near Christie's natal Jupiter.

On 21 March Wilk discovered long period Comet Wilk (C1930 F1) at 27 Aries, which reached perihelion at 28 Aries where it became visible. It moved on to align with Venus at 25 Aries, Mercury at 24 Aries and the Sun at 21 Aries, all points close to Christie's IC and opposing her natal Mercury.

Finally Comet Schwassmann-Wachmann 3 (73P) was discovered on 27 April at 12 Scorpio, close to her natal Venus.

Once again we can see the base of Christie's chart being activated, along with her Jupiter (travel) and Venus (relationships).

The Mousetrap

Christie's prolific writing extended to stage plays, the most famous being *The Mousetrap*, which has been continuously performed at London's West End since its premier on 6 October 1952 (aside from a year-long break during the Covid 19 pandemic from March 2020 to May 2021). It has been performed more than 30,000 times and been attended by more than 10 million people making it by far the longest running show in the world.

The play was originally written for radio as a birthday gift for Queen Mary, the consort of King George V, and was broadcast under the title *Three Blind Mice* on 30 May 1947. That day transiting Neptune was conjunct Christie's natal Moon at 8 Libra, both Pluto and Saturn were squaring natal Venus ruling her 5th house (the stage) and her MC. Furthermore transiting Saturn was conjunct her Chiron, transiting Uranus and Mercury were very close to her Descendant, and the Sun had just passed her Neptune-Pluto conjunction. Finally both Venus and Mars opposed natal Venus. (Figure 6.7).

On top of all that we have two comets adding their influence to this pivotal moment in Christie's creative life. Comet Pajdusakova-Robert-Weber was discovered on 30 May 1946, becoming visible for just four days from 1 to 4 June. During this very short period it travelled across an incredible 90 degrees of the zodiac aligning with the transiting South Node at 20 Sagittarius near Christie's Ascendant and her Mars on 2 June. It had reached 25 Libra beside her MC by the time it disappeared from view two days later.

Comet Rondanina-Bester (C/1947 F1) was visible from 12 April 1947 and four days later aligned with transiting Mars at 3 Aries opposite Christie's Moon. On 28 April it aligned with transiting Mercury at 19 Aries opposite natal Mercury and Uranus, and on 21 May it aligned with transiting Venus at 1 Taurus, crossed the celestial equator at 3 Taurus to square natal Jupiter. Finally on 7 June, four days before it disappeared, the comet aligned with transiting Mars at 12 Taurus opposite natal Venus.

When *The Mousetrap* premiered at the Theatre Royal in Nottingham on 6 October 1952, transiting Neptune and Mercury were both at 21 Libra conjunct Christie's Uranus. Saturn was at 17 Libra and the Sun was at 13 Libra, the pair sandwiching natal Mercury from two degrees on either side. In addition Venus at 10 Scorpio was conjunct natal Venus and Mars at 25 Sagittarius was conjunct natal Mars and her Ascendant.

On top of all that planetary activation, Comet Schaumasse (24P) - who we met around the time Christie met her first husband after it was discovered on 1 December 1911 at 14 Libra (conjunct natal Mercury) - made one of its seven yearly reappearances in the sky crossing natal Chiron and Saturn on its path. And in 1952 it was the brightest it had ever been!

Later life

Christie suffered a heart attack in 1974, the year of her Uranus return, which brought a final end to her writing career. Her final novel *Postern of Fate* had been published in October 1973, but two years later the previously conceived novel *Curtain: Poirot's Last Case* was published with *The New York Times* running an obituary for the fictional detective and featuring the book's cover on its front page.

During those few years transiting Pluto crossed Christie's Moon, while Neptune opposed that natal Neptune-Pluto conjunction.

Comet Kohoutek (C/1973 E1) was discovered on 30 March 1973 at 3 Leo, close to her natal Chiron and opposite natal chart ruler Jupiter. On 11 November it aligned with Pluto at 5 Libra, one degree from her natal Moon, and on 28 November it aligned with Uranus at 25 Libra one degree from her MC. To cap it all, it aligned with Mercury at 22 Sagittarius on Christmas Eve right next to Christie's Ascendant and natal Mars.

The Queen of Crime, as she was fondly known, died on 12 January 1976 while Saturn transited retrograde across her natal Chiron opposite

Comets and Natal Charts   49

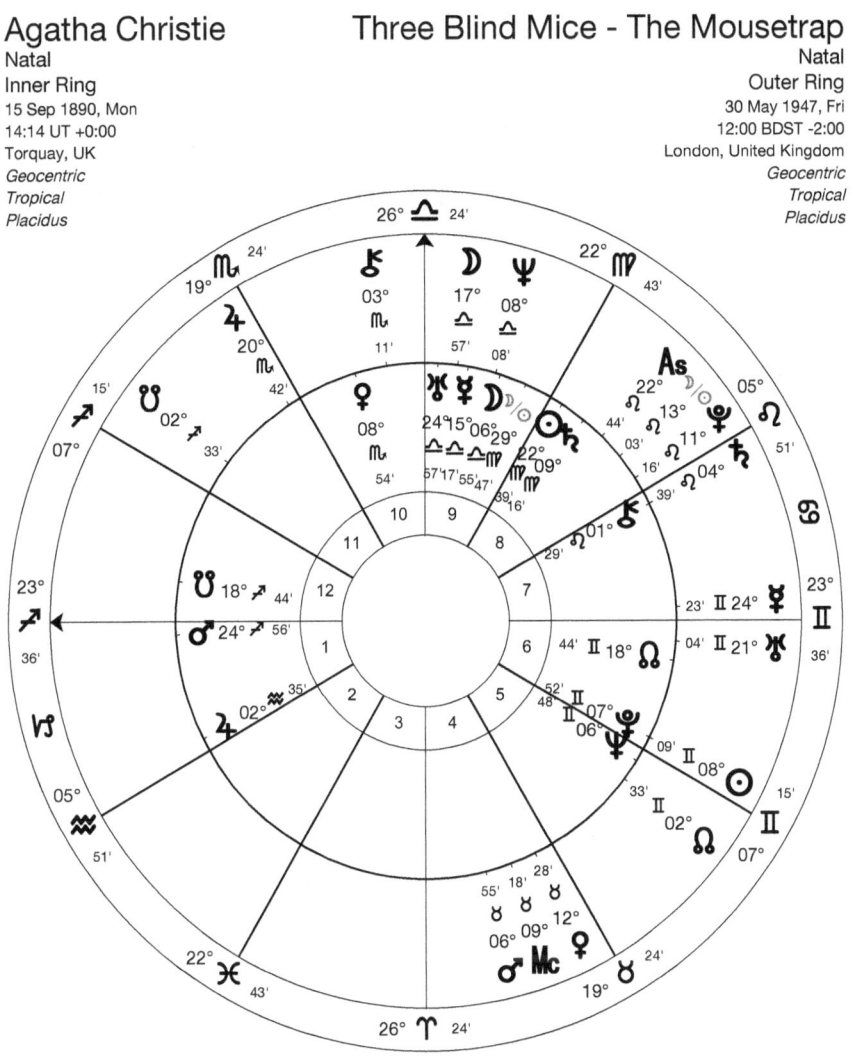

Figure 6.7

Jupiter, while Uranus approached a conjunction with natal Venus and Chiron was near her IC.

Allowing comets to have the last word on her life story, Comet Suzuki-Saigusa-Mori (C/1975 T2) appeared in the sky between October and November 1975. On 29 October it aligned with transiting Mercury at 17 Libra within two degrees of natal Mercury. The following day it reached 6 Scorpio where it aligned with the Sun, two degrees from the ruler of her Midheaven, Venus.

Chapter 7

How to find comets using the JPL Horizons system

The greatest challenge for astrologers wanting to study comets will be finding them in a chart. Comets are by their nature unpredictable and have unique orbits, which means they frequently change their trajectory. This makes it impossible for our commonly used astrology software to calculate their position. However we can obtain accurate information from the NASA Jet Propulsion Laboratory's Horizons On-Line Ephemeris System.

The JPL Horizons System is an amazing online service giving us access to a wealth of high precision data about objects in our solar system, including almost 1.5 million asteroids, around 4,000 comets and 300 planetary satellites, all provided by JPL's Solar System Dynamics Group.

To be able to use the ephemeris effectively you will require the full name and number of the comet in question, which will be available on the internet. Many comets carry the same name, but each will have its unique number. For example, there are dozens of Comet Leonards, but the one that appeared at the end of 2021 is known as C/2021 A1 (Leonard).

To find a comet:

1. Search for JPL Horizons System, or use the link: https://ssd.jpl.nasa.gov/horizons/app.html#/

2. You will see five fields and a green 'Generate Ephemeris' button at the bottom of the screen

3. The first field 'Ephemeris Type' can be ignored

4. In the second field 'Target Body', click on the blue 'Edit' button, which asks you to 'Specify the Target Body'

5. In the 'Lookup the Specified Body' search box, type in the name of the comet, eg. C/2021 A1 and click the green search button. Make sure you have the correct name for your target.

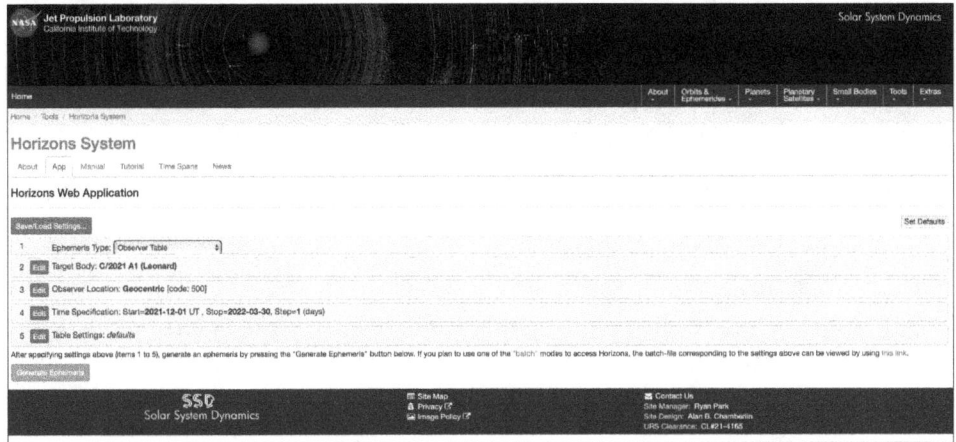

6. Having returned to the main page, go to the fourth field 'Time Specification', click the blue 'Edit' button and enter the period of time you are interested in by clicking on 'Use Specified Time Span'.

7. Returning to the main page again, we go to the fifth field 'Table Settings', which is a critical setting for astrologers. First click on the blue 'Edit' button. Then:
   - select No.31 'Observer ecliptic lon. & lat.', which will provide the comet's all-important degree on the ecliptic (though remember that comets, unlike planets, do not stay on the ecliptic);
   - if you require the comet's declination then make sure No.1 'Astrometric RA & DEC' is selected;
   - to find out what constellation (as opposed to tropical zodiac sign) the comet is passing across, then click on No.29 'Constellation ID' (for more information on constellation names and abbreviations, click on https://www.aavso.org/constellation-names-and-abbreviations;
   - then deselect all the other fields.
   - finally click on the blue 'Use Specified Settings' button at the bottom of the page, which will take you back to the main page.

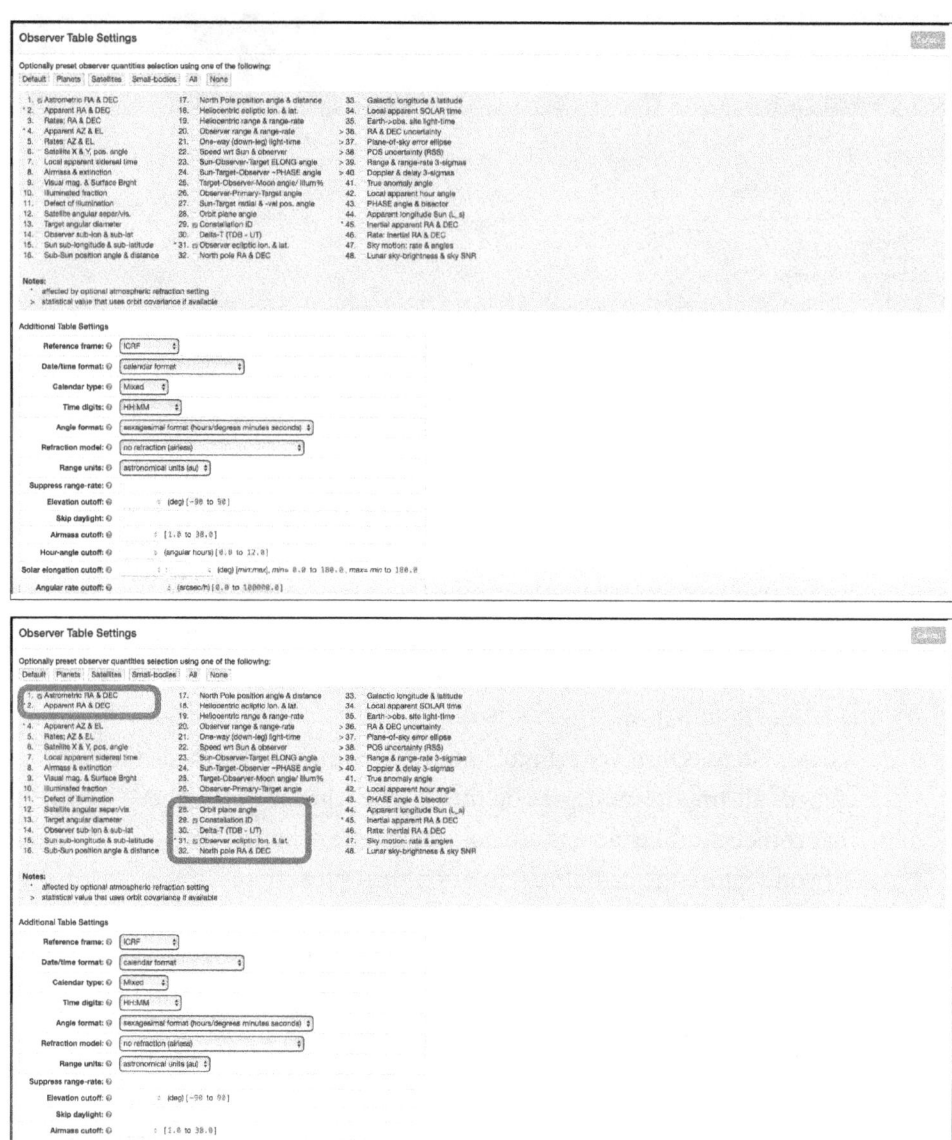

8. Press the green 'Generate Ephemeris' button at the foot of the page. This takes you to a page with a lot of data. Halfway down the page there are several columns, the first being a list of dates. The fifth column along is marked 'ObsEcLon', which shows the comet's degree on the ecliptic. The longitude is expressed somewhere

between 0 and 360 degrees, so 0 degrees is 0 Aries, and 359 is 29 Pisces.

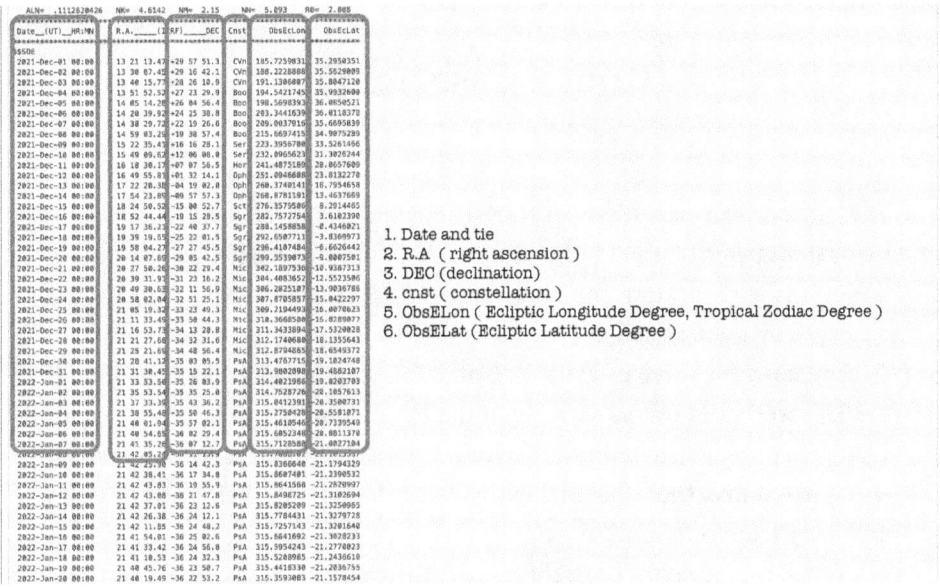

1. Date and tie
2. R.A (right ascension)
3. DEC (declination)
4. cnst (constellation)
5. ObsELon (Ecliptic Longitude Degree, Tropical Zodiac Degree)
6. ObsELat (Ecliptic Latitude Degree)

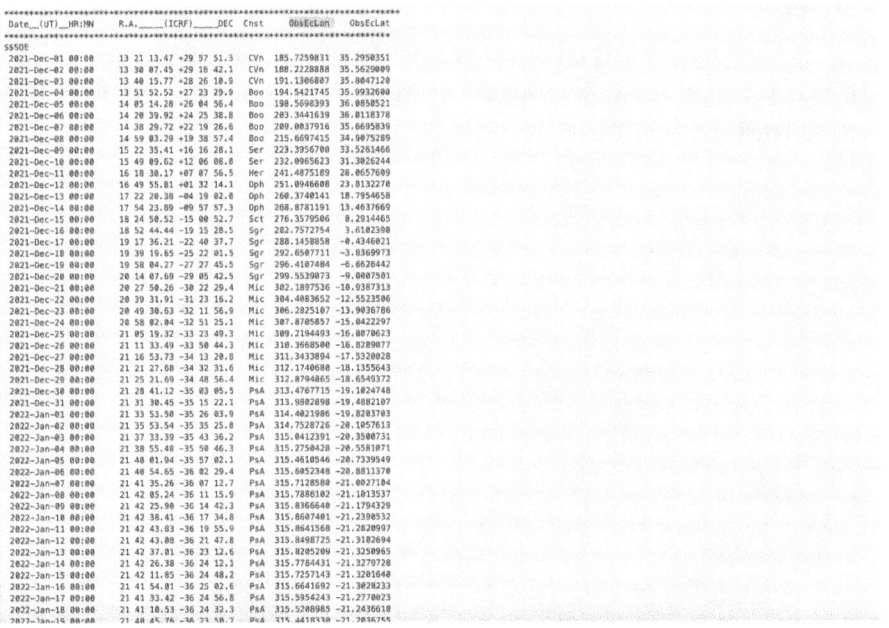

9. The sixth column marked 'ObsEcLat' shows the comet's elliptical latitude, a coordinate rarely used by astrologers.

Chapter 8
Comet Case Studies

The ancient astrologers believed that comets appeared within a year of natural disasters of one kind or another. I tested their belief by examining the events surrounding the appearance of 15 comets close enough to be observed by the naked eye.

I researched three of Halley's Comet's appearances in 1066, 1910 and 1986, as well as 14 others starting in the 16th century up to the present day. I looked at the constellations, the zodiac signs, the degrees through which they passed and the aspects they made to planets on their path towards and away from our Sun.

I took into account both the sign in the tropical zodiac and the actual constellation as these differ, and I allowed a maximum five degree orb for aspects.

While I realise this represents a fraction of the number of comets that have been seen in the sky over the centuries, I hope my observations can stimulate further research into this relatively unexplored subject.

Halley's Comet 1066 (1P/Halley)

Type of Comet: periodic

Orbital period: 72–80 years

Astronomical data

	Date	Position
First seen	2 Apr 1066	09 Aquarius
Perihelion	20 Mar 1066	11 Pisces
Crossed Ecliptic	25 Apr 1066 02:50 UT	00 Cancer

Tropical Zodiac

Date	Sign
20 Mar - 18 Apr	Pisces
18 Apr - 22 Apr	Aries
22 Apr - 23 Apr	Taurus
23 Apr - 25 Apr	Gemini
25 Apr - 28 Apr	Cancer
28 Apr - 24 May	Leo

Constellations

Note: not to be confused with the 12 signs of the tropical zodiac.

Date	Constellation
20 Mar - 14 Apr	Pisces
14 Apr - 16 Apr	Pegasus
16 Apr - 21 Apr	Pisces
21 Apr - 22 Apr	Aries
22 Apr - 24 Apr	Taurus
24 Apr - 26 Apr	Gemini
26 Apr - 29 Apr	Cancer
29 Apr - 02 May	Leo
02 May - 24 May	Sextans

Planetary alignments

Date	Planet	Zodiac location
09 Apr 1066	Venus	11 Pisces
19 Apr 1066	South Node	4 Aries
22 Apr 1066	Mercury	2 Taurus
22 Apr 1066	Sun	7 Taurus
24 Apr 1066	Neptune	21 Taurus
28 Apr 1066	Jupiter	29 Cancer

Perhaps the most famous comet sighting in English history came in the fateful year of 1066, a date emblazoned on every school child's memory as the year of the Norman Conquest that ended Anglo-Saxon rule of the country.

The first report of the comet was on 2 April after it had already reached perihelion on 20 March at 11 Pisces in the tropical zodiac. Historical records suggest it appeared to be four times the size of Venus and shone as bright as a quarter of the Moon, coming within 0.10AU of Earth. The Irish *Annals of the Four Masters* described the comet as: "A star that appeared on the seventh of the Calends of May, on Tuesday after Little Easter. It was brighter than the moon, and it was visible to all for four nights." This is contradicted by modern astronomical software, which calculates Halley's Comet reached its brightest magnitude on 24 April and would have been visible until the end of May.

Located in the tropical sign as well as the constellation of Pisces, one would expect the comet's appearance to coincide with disasters related to water, such as floods or shipwrecks (the Normans invaded by sea), as well as the death of kings.

Events

In 1066 when Halley's Comet was seen in the sky above England it was viewed as an omen, as it indeed turned out to be. The country faced invasions from the north and the south that led to King Harold II's death at the Battle of Hastings on 14 October. William, Duke of Normandy, claimed the English throne and ended the reign of the Anglo-Saxons once and for all, going down in history as William the Conqueror.

The comet's appearance was immortalised in the famous Bayeux Tapestry that recorded the events leading up to the famous battle in which it is described as a star. It is also mentioned in the Italo-Byzantine chronicle of Lupus Protospatharios as a "comet-star" appearing in 1067, though he was out by a year. He records the victory of William, though he erroneously calls him Robert, and he also mentions the death of the Byzantine emperor Constantine Ducas in 1067.[67]

Tapestry of Bayeux (Normandy) with Halley's comet. Text reads ISTI MIRANT STELLA: "These (people) are looking in wonder at the star."

The year 1066 also saw the death of King Stenkil of Sweden, which triggered a violent civil war between Christians and pagans, and it was the year when the young Emperor Yingzong of China fell ill, leading to his death the following January.[68]

The Battle of Hastings overshadowed other major historical events of the time because it played such a major role in the history of the English-speaking world, but there was no shortage of other disruptive events in the region. The battle itself was preceded by the Norse invasion in the north of

England earlier in the year and in 1067 there were rebellions against the Norman conquerors.

In 1067 medieval France saw a major (but unsuccessful) uprising by serfs in Viry against their lords and the canons of the church of Notre Dame de Paris over issues including the right to marry the woman of their choice without church approval – we can note here that Halley's Comet aligned with Jupiter in Cancer, denoting the church and powerful people, and Venus which is associated with marriage.[69]

When Halley's Comet crossed paths with Jupiter in April 1066, the planet was at 29 Cancer whose symbolism includes the 'mother church'. A number of other religious tensions came to a head around this time, including:

- the excommunication of Milan's Archbishop Guido da Velate by Pope Alexander II;
- the rise of the Pataria movement, and the martyrdom and subsequent canonisation of its leader Ariald; and
- Canterbury Cathedral being burnt to the ground.[70]

On 30 December 1066 there was a massacre of Jews in the Spanish city of Granada, which was under the rule of the Ottoman empire. A Muslim mob stormed the royal palace, killed the Jewish vizier Joseph ibn Naghrela who was suspected of poisoning the king's son, and then went on a rampage killing Jews living in the city, with the survivors dispersing to other towns. I would associate this event with the comet's path through Aries and Leo.

There is speculation that the comet was observed in China on the morning of 2 April 1066 in the constellations of Pisces and Pegasus, with a length of about 7 Chi (about 160cm today).[71]

Examining records of Chinese court history in the 1084 volume written by Sima Guang entitled *Compiled by Comprehensive Mirror in Aid of Governance* (續資治通鑑) it appears that the somewhat sickly Emperor Yingzong of Song considered the comet to be a bad omen and decided to go on a fast, refusing ministers' advice to return to a normal diet. He also requested court historian Sima Guang to start compiling a record of events covering the previous 15 centuries.

Reading this document we learn that the emperor believed the comet was warning him to strengthen the country's border defences. We also learn there was a plague of locusts in 1066 and a drought was followed by floods in Shaanxi province causing a poor harvest. The following winter saw several major earthquakes, severe cold weather and an invasion of refugees into the capital.

Emperor Yingzong died in 1067 after a short and controversial reign of just four years having always suffered from poor physical and mental health.

Looking at these events through an astrological eye, we might consider the severe cold and floods to relate to the comet passing through the water signs of Pisces and Cancer, wars to relate to Aries and Leo, and the food shortages resulting from droughts and locusts as well as earthquakes relating to the sign of Taurus. I would also relate the compilation of important historical documents, the air-borne locust plague and the refugee crisis to Gemini and Mercury.

The Great Comet of 1577 (C/1577 V1)

Type of Comet: non-periodic

Astronomical data

	Date	Position
Discovery	1 Nov 1577	
Last perihelion	27 Oct 1577	
Last seen	26 Jan 1578	21 Pisces
Crossed celestial equator	22 Nov 1577 00:20	00 Aquarius
Crossed ecliptic	08 Nov 1577 18:03	18 Sagittarius

Tropical Zodiac

Date	Sign
01 Nov 1577 - 03 Nov 1577	Scorpio
03 Nov 1577 - 11 Nov 1577	Sagittarius
11 Nov 1577 - 22 Nov 1577	Capricorn
22 Nov 1577 - 15 Dec 1577	Aquarius
15 Dec 1577 - 25 Jan 1578	Pisces

Constellations

Note: not to be confused with the 12 signs on the tropical zodiac.

Date	Constellation
01 Nov 1577 - 04 Nov 1577	Lupus
04 Nov 1577 - 06 Nov 1577	Scorpius
06 Nov 1577 - 09 Nov 1577	Ophiuchus
09 Nov 1577 - 12 Nov 1577	Sagittarius
12 Nov 1577 - 14 Nov 1577	Scutum
14 Nov 1577 - 24 Nov 1577	Aquila
24 Nov 1577 - 29 Nov 1577	Delphinus
29 Nov 1577 - 07 Dec 1577	Equuleus
07 Dec 1577 - 26 Jan 1578	Pegasus

Fixed Star conjunctions

Date	Star
04 Nov 1577	Antares
20 Jan 1578	Scheat

Planetary alignments

Date	Planet	Zodiac location
07 Nov 1577	Mercury	14 Sagittarius
14 Nov 1577	Saturn	10 Capricorn
22 Nov 1577	Uranus	00 Aquarius

Events

The Great Comet of 1577 stands out in history for being observed and commented upon right across the world.

From a stargazer's point of view, perhaps of most significance are the efforts of Danish astronomer Tycho Brahe whose detailed observations of this comet enabled him to conclude its orbit was beyond that of the Moon, contradicting Galileo who later claimed comets were "an optical phenomenon" after viewing them through his telescopes. It was Brahe's detailed records of the 1577 comet that helped his student Johannes Kepler to arrive at his laws of planetary motion.[72] We can note here that the about-to-be discovered planet Uranus was just entering Aquarius, the planet and sign of scientific breakthroughs, when the comet passed by. It also crossed Mercury at 14 Sagittarius right on Brahe's MC. (Figure 8.2.1).

Queen Elizabeth I famously defied her advisers who warned her not to look upon this comet as it was considered a bad omen for royalty to do so.[73] The comet passed by Saturn in Capricorn, reflecting the major political changes of the time. This was the 'Age of Discovery' and during the comet's appearance the Queen gave her backing to the English privateer Martin Frobisher's second expedition in search of the North West Passage, and to Sir Francis Drake crossing the Strait of Magellan at the southern tip of South America for the first time in 1577. We can note the comet passed through the adventurous constellations of Aquila and Delphinus as well as the zodiacal water signs of Scorpio and Pisces, and very close to fixed star Scheat in the constellation Pegasus.

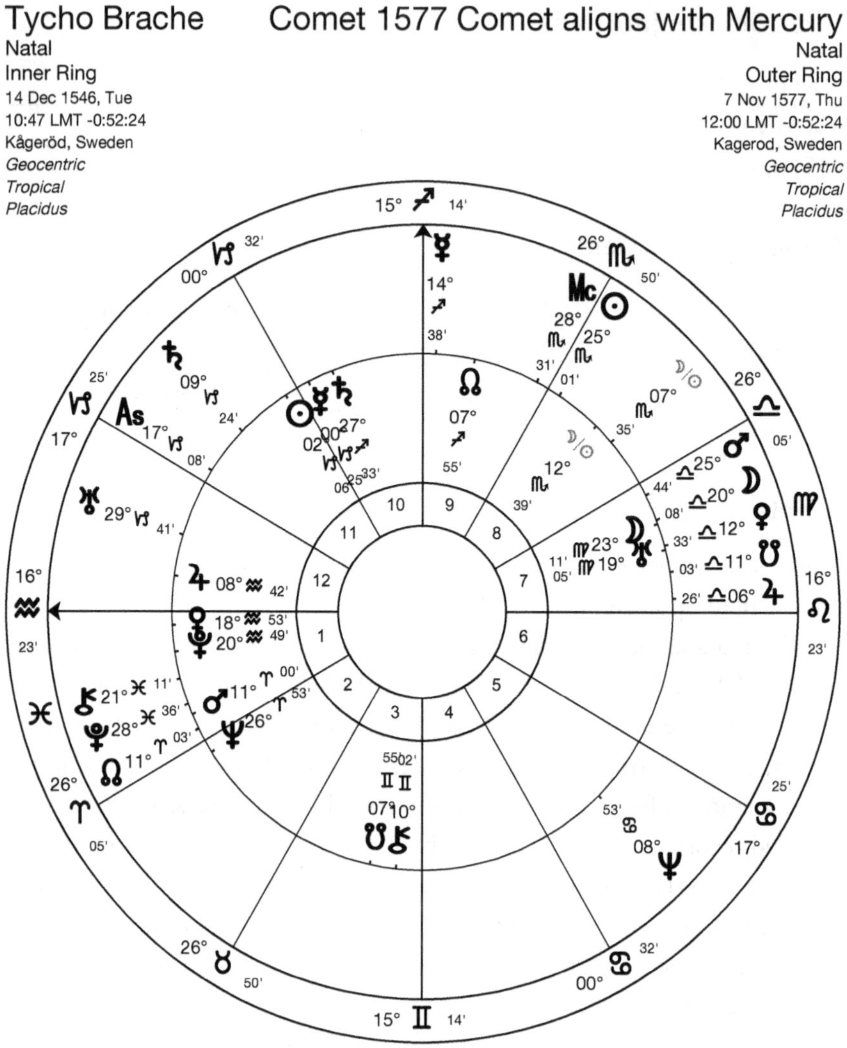

Tycho Brache Comet 1577 Comet aligns with Mercury
Natal   Natal
Inner Ring   Outer Ring
14 Dec 1546, Tue   7 Nov 1577, Thu
10:47 LMT -0:52:24   12:00 LMT -0:52:24
Kågeröd, Sweden   Kågerod, Sweden
Geocentric   Geocentric
Tropical   Tropical
Placidus   Placidus

Figure 8.2.1

At the time Europe was going through the 80 year long Dutch War of Independence against Spain, one of the most significant conflicts the continent has been through. There was also fighting between Catholics and Huguenots in southern France.[74] Also warring against the Spanish was Sultan Murad III, who held the comet responsible for the plague that struck the Ottoman Empire at the time. These events can be related to the comet's passage through the constellations Sagittarius, Aquila and Scutum.

Meanwhile on the other side of the Eurasian landmass, the title Dalai Lama was used for the first time in 1578. The story goes that the

Tibetan Buddhist spiritual leader Sonam Gyatso travelled 1,500 miles to Mongolia at the invitation of the Mongol chief Altan Khan. It was the Khan who bestowed on him the title Dalai, which is the Mongolian translation of "gyatso", which means 'ocean'. The Mongol leader went on to make Buddhism the state religion and reform the country's law accordingly. From this time on, Buddhism spread rapidly to replace the nation's shamanic traditions. We can see here the influence of the signs of Sagittarius and Pisces which the comet passed through on its journey across the sky while visible to the naked eye.

Meanwhile in China the *History of the Ming Dynasty* records this dramatic comet being observed on 14 November 1577 in the south west, describing it as pale in colour, ten metres long and with clouds (tails) like a white rainbow.[75] It disappeared after one month having passed through Wei Xu (the tail end of the Scorpio constellation), Jisu (the bow and arrow of the constellation Sagittarius), Duo Xu (the body of Sagittarius), Nio Xu (Capricorn and Aquila) and Nu Xu (Aquarius).

In 1578 the Chinese physician Li Shizhen completed the *Great Pharmacopoeia* (or *Compendium of Materia Medica*), a classic of traditional Chinese medicine that classified plants according to their medical properties.[76] Li Shizhen was born on 3 July 1518, and the comet crossed the ecliptic on his South Node at 18 Sagittarius, also aligning with transiting Uranus and his natal Chiron, symbolising the wounded healer, at 0 Aquarius. The comet also passed through the constellation Ophiuchus, representing Asclepius, the god of healing.

Important people who passed away this year include:

- Catherine of Austria, Queen of Portugal
- Sebastiano Venier, Doge of Venice
- Louis I, Cardinal of Guise, French cardinal
- James Hepburn, 4th Earl of Bothwell, consort of Mary, Queen of Scots
- Uesugi Kenshin, Japanese samurai and warlord
- Thomas Doughty, English explorer
- Sebastian of Portugal
- Abu Marwan Abd al-Malik I Saadi, King of Morocco
- Abu Abdallah Mohammed II Saadi, King of Morocco
- Gonzalo II Fernández de Córdoba, Governor of the Duchy of Milan

Great Comet of 1677 (C/1677 H1)

Type of Comet: non periodic

Astronomical data

	Date	Position
Discovery	27 Apr 1677	00 Taurus
Perihelion	6 May 1677	19 Taurus
Crossed celestial equator		
Crossed ecliptic	19 May 1677 10:40 UT	28 Taurus

Tropical Zodiac

Date	Sign
27 Apr 1677 - 23 May 1677	Taurus
23 May 1677 - 30 May 1677	Gemini

Constellations

Note: not to be confused with the 12 signs of the tropical zodiac.

Date	Constellation
27 Apr 1677 - 03 May 1677	Triangulum
03 May 1677 - 08 May 1677	Perseus
08 May 1677 - 09 May 1677	Aries
09 May 1677 - 30 May 1677	Taurus

Planetary alignments

Date	Planet	Zodiac location
02 May 1677	Sun	02 Taurus
22 May 1677	Saturn	29 Taurus

Events

The earliest record I have come across for this comet comes from Polish astrologer Johannes Hevelius who spotted it on 27 April.[77] English astrologer William Lilly made a prediction based on its appearance in his astrological judgment for the year 1678.[78] The German astronomer Johan

Voigt published an engraving of the comet[79] and Danish astronomer Ole Rømer noted its passing.

The period of 1677-78 produced some fascinating discoveries. The Dutch textile merchant Antonie van Leeuwenhoek became one of the world's first microbiologists using microscopes to discover spermatozoa.[80] Could this be linked to the comet's passage through the constellation Perseus, related to primal sexual energy?

The foundational textbook of mathematics known as *Cocker's Arithmetik*, by English engraver and teacher Edward Cocker, was published posthumously in 1678 and went on to become the standard work in grammar schools for the next 150 years. Could this relate to the comet's passage through the constellation Triangularum, which is linked to mathematics, geometry and a wide range of scientific subjects?

This brief period also saw the publication of:

- the first English star atlas by royal hydrographer John Seller;
- one of the first theories of light travelling in waves by Dutch mathematician and physicist Christian Huygens;
- English polymath Robert Hooke's discovery of the fundamental laws of elasticity that led to him claiming to have invented the first portable timepiece, namely the watch;[81]
- Robert Plot's influential *Natural History of Oxfordshire*, which included the first known illustration of a dinosaur bone, thought at the time to be that of a giant. The book is credited with inspiring Elias Ashmole to create Britain's first public museum in 1678, the Ashmolean, which is still open today.

William Lilly predicted the comet's passage through Taurus would impact countries ruled by the sign of the Bull, specifically mentioning Russia, Poland, Norway, Italy and Ireland. In 1677 the Russian-Ukrainian cavalry successfully re-captured the Ukrainian city of Chyryryn from an invading force of the Turkish-Crimean army, which lost 20,000 men in the battle.[82]

The passage through Taurus could be seen to relate to the signing of the Statute of Frauds into English law, that concerns issues of contracts, finance and land.[83]

Great Comet of 1844 (C/1844 Y1)

Type of Comet: non periodic

Astronomical data

	Date	Position
Discovery	18 Dec 1844	07 Capricorn
First seen	18 Dec 1844	07 Capricorn
Perihelion	14 Dec 1844	29 Scorpio
Crossed ecliptic	14 Dec 1844	29 Scorpio

Tropical Zodiac

Date	Sign
18 Dec 1844 - 29 Dec 1844	Capricorn
29 Dec 1844 - 11 Jan 1845	Aquarius
11 Jan 1845 - 24 Jan 1845	Pisces
24 Jan 1845 - 30 Jan 1845	Aries

Constellations

Note: not to be confused with the 12 signs of the tropical zodiac.

Date	Constellation
18 Dec 1844 - 27 Dec 1844	Sagittarius
27 Dec 1844 - 01 Jan 1845	Microscopium
01 Jan 1845 - 11 Jan 1845	Grus
11 Jan 1845 - 16 Jan 1845	Phoenix
16 Jan 1845 - 28 Jan 1845	Sculptor
28 Jan 1845 - 30 Jan 1845	Fornax

Planetary alignments

Date	Planet	Zodiac location
27 Dec 1844	Mercury	25 Capricorn
01 Jan 1845	Saturn	06 Aquarius
08 Jan 1845	Neptune	22 Aquarius
24 Jan 1845	Jupiter	00 Aries
26 Jan 1845	Uranus	03 Aries

Events

The Great Comet of 1844 was visible for just over one month, but it followed two comets that appeared the previous year – the Great Comet of 1843 (C/1843 D1) and Comet Mauvais (C/1843 N1) – which will no doubt have contributed to the upheaval of this period. My attention was drawn to this particular comet due to astronomers pointing out its behaviour resembled that of Comet Atlas (C/2019 Y4).[84]

On 21 December 1844, just three days after the comet was first spotted in the sky, the Rochdale Pioneers gave birth to the modern day co-operative movement drawing up the principles that still underpin the way co-ops operate around the world today.[85] This would relate to the comet aligning with Mercury, the planet of trade and commerce, in Capricorn, the sign of structure, and also with Saturn and Neptune in Aquarius, representing the consolidation of utopian social ideals.

The comet's passage across the heavens coincided with a critical moment in the history of the United States – the annexation of the state of Texas, which had been part of the republic of Mexico.[86] Texas declared its independence from Mexico on 2 March 1836. Using the state's midnight chart (Figure 8.4.1), the Great Comet of 1844 crossed the ecliptic and was at perihelion simultaneously at 29 Scorpio, conjunct the Texan Ascendant at 0 Sagittarius. It aligned with Saturn at 22 Aquarius next to the Mars-Mercury conjunction in the Texan chart and was discovered at 7 Capricorn opposite the Texan Jupiter. The annexation had long term consequences, not least the two year Mexican-American War from April 1846 to February 1848 that led to the establishment of the USA's southern border along the Rio Grande and the country acquiring the current states of California, Nevada, Utah, most of New Mexico, Arizona, Colorado, parts of Texas, Oklahoma, Kansas and Wyoming.[87]

Many striking events occurred in 1845 that appear to bear the hallmark of the comet's astrological signification,[88] especially its discovery and passage through the earth sign of Capricorn, related to land, structure and food:

- In April a major earthquake devastated Mexico City and surrounding areas;

Republic of Texas

Natal
2 Mar 1836, Wed
00:00 LMT +6:22:34
West Columbia, Texas
Geocentric
Tropical
Placidus

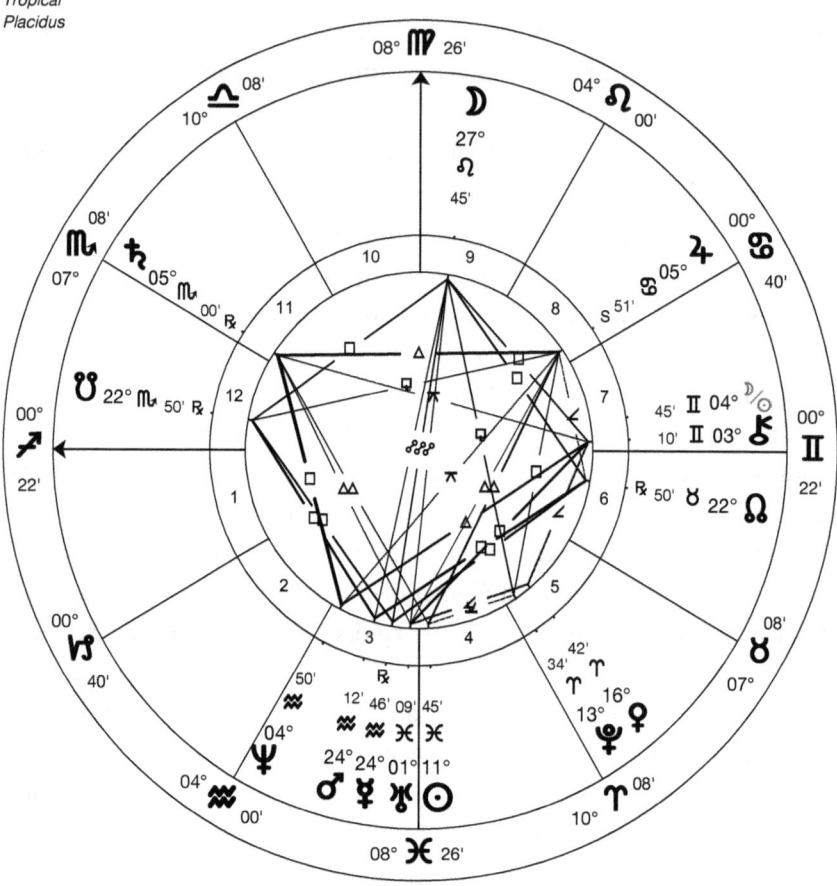

Figure 8.4.1

- In May a suspension bridge collapsed in the English town of Great Yarmouth killing 80 people, mostly children;

- In August a tornado destroyed two factories and killed 200 people in France;

- In September the first appearance of potato blight led to the Great Famine in Ireland.

Other events during this period that strike me as bearing some relation to the comet include:

- The foundation of Baylor University, the oldest university in the state of Texas, and the three Queen's Colleges in Belfast, Cork and Galway, Ireland (Sagittarius constellation);
- The rise of the anti-slavery movement in America (Neptune in Aquarius);
- The first use of anaesthesia in childbirth (Neptune in Aquarius);
- The completion of the longest railway bridge in the world stretching two miles across the Venetian Lagoon (Mercury in Capricorn).

Great Comet of 1901 (C/1901 G1)

Type of Comet: non periodic

Astronomical data

	Date	Position
Discovery	12 Apr 1901 pre-dawn	12 Aries
Last Perihelion	24 April 1901	19 Aries
Last day to be seen	23 May 1901	03 Cancer
Crosses celestial equator	04 May 1901 19:57 UT	26 Taurus
Crosses Ecliptic	16 Apr 1901	12 Aries

Tropical Zodiac

Date	Sign
12 Apr 1901 - 27 Apr 1901	Aries
27 Apr 1901 - 05 May 1901	Taurus
05 May 1901 - 20 May 1901	Gemini
20 May 1901 - 23 May 1901	Cancer

Constellations

Note: not to be confused with the 12 signs of the tropical zodiac.

Date	Constellation
12 Apr 1901 - 26 Apr 1901	Pisces
26 Apr 1901 - 02 May 1901	Cetus
02 May 1901 - 03 May 1901	Taurus
03 May 1901 - 06 May 1901	Eridanus
06 May 1901 - 08 May 1901	Taurus
08 May 1901 - 23 May 1901	Orion

Planetary alignments

Date	Planet	Zodiac location
30 Apr 1901	Venus	9 Taurus
30 Apr 1901	Sun	9 Taurus
04 May 1901	South Node	22 Taurus
12 May 1901	Pluto	16 Gemini
18 May 1901	Neptune	27 Gemini

Events

One of the most significant societal shifts of the period at the beginning of the 20th century that can be linked to the Great Comet of 1901 relates to its alignment with the Sun-Venus conjunction at 9 Taurus on 30 April. At this time the women's suffrage movement was really gaining momentum and by the following year on 12 February 1902 the First Conference of the International Woman Suffrage Alliance was held in Washington DC, a key moment in the history of the movement.[89]

The Australian federation was founded at midnight at the start of the year 1901 when the Moon was at 12 Taurus, very close to where the comet passed the Sun-Venus conjunction on 30 April. In the following June, Australia became the first independent country to grant women the right to vote and stand for public office at a national level, setting a precedent soon to be followed by other nations.[90]

This Great Comet first appeared in the sign of Aries, which one would expect to coincide with disasters associated with the element of fire. These occurred the following year in profusion, with four major volcanic eruptions in the Caribbean region:[91]

- The Soufrière volcano erupted on the island of St Vincent on 6 May killing 1,680 people, injuring 600 and leaving 4,000 homeless;
- Within hours Mount Pelée on the island of Martinique exploded taking the lives of 29,000, and with it the last remnants of the Carib culture;
- Mount Pelée erupted again on 30 August, killing a further 1,000 people and destroying the town of Le Mourne-Rouge;
- October saw one of the 20th century's largest eruptions at Santa Maria, in Guatemala, causing the death of more than 6,000 people.

Even more notable from an astrological perspective are the correlations with the comet's passage through the sign of Taurus, with its connection to the material and money. Three days after the comet left the sign of the Bull, the New York Stock Exchange (NYSE) crashed for the very first time in what is known as the Panic of 1901.[92] The crash occurred at lunchtime on 8 May, when the Moon, Jupiter and Saturn were transiting the NYSE Descendant at 13 Capricorn, squaring the point where the comet was first

72 Comets in Astrology

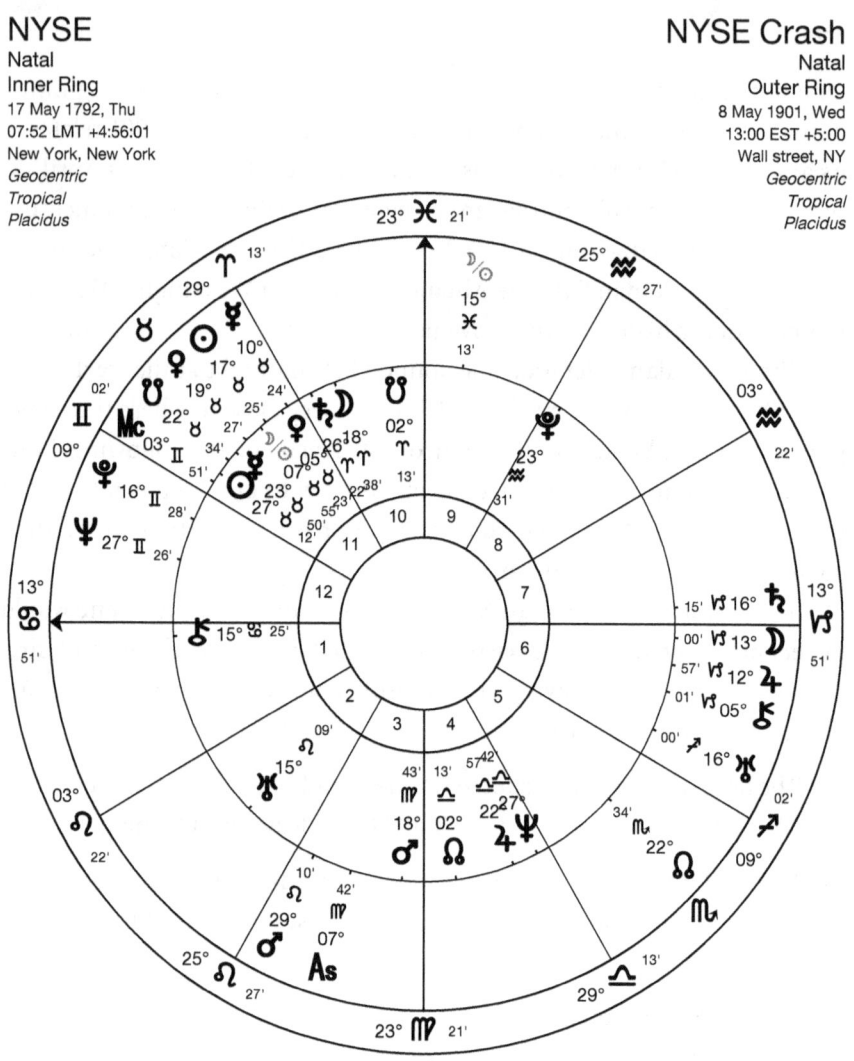

Figure 8.5.1

seen and where it crossed the ecliptic at 12 Aries. Other astrological significations include transiting Uranus at 16 Sagittarius squaring the NYSE Mars at 18 Virgo. On 4 May the comet aligned with the South Node, the symbol of loss and reduction, at 22 Taurus near NYSE's Mercury, and then crossed the celestial ecliptic at 26 Taurus near NYSE's Sun. (Figure 8.5.1)

If we apply the ancient syzygy technique used by Tycho Brahe, we can see that on the Full Moon on 3 May 1901, the comet was conjunct Venus, the ruler of the second house, at 13 Taurus, opposite the Moon and moving towards the South Node, the symbol of loss and reduction.

The element of earth is reflected in several catastrophes during the two years in question:

- March 1901 saw an extreme 7.2Mw earthquake in the Black Sea off the north east coast of Bulgaria, the most powerful this area has ever seen, that triggered a devastating tsunami affecting the province of Dobrich;
- In May, 81 miners were killed in an accident at the Universal Colliery in South Wales[93];
- April 1902 saw Guatemala experience a severe 7.5Mw earthquake that killed an estimated 800 to 2,000 people;
- In July that year 112 miners died in the Rolling Mill Mine disaster in Pennsylvania;
- The same month the famous tower of St Marks Campanile in Venice collapsed;
- The following month, 100 miners died in a pit explosion in Wollongong, Australia;
- Less than a month later a 7.7Mw earthquake killed 10,000 people in China and Kyrgistan.[94]

Breakthroughs, discoveries and significant events during this period include:

- The world's first passenger-carrying trolleybus going into service in Biela Valley, Germany (Gemini);
- Guglielmo Marconi received the first transatlantic radio signal sent from Poldhu, in Cornwall, England, to St John's, in Newfoundland, Canada, in December 1901 (Gemini);
- In July 1901 the first UK fingerprint bureau was set up by Edward Henry at Scotland Yard, the headquarters of London's Metropolitan Police;[95]
- In medicine Dr Alois Alzheimer diagnosed the first case of the neurodegenerative disease that bears his name;[96]

- New Zealand became the first country to require state registration of nurses;[97]
- The world's first cinema, the Electric Theatre, opened in Los Angeles;
- The first teddy bear was created, inspired by President "Teddy" Roosevelt refusing to shoot a bear cub in 1902.[98]

Finally, on 6 September 1901 the US president William McKinley was assassinated in New York by an anarchist. The last sighting of Comet 1901 was at 19 Aries, right on McKinley's Pluto.[99]

Great Comets of 1910

The Great Daylight Comet of 1910 (C/1910 A1)

Type of Comet: non periodic

Astronomical data

	Date	Position
Discovery	12 Jan 1910	10 Capricorn
Perihelion	17 January 1910	23 Capricorn
Crossed celestial equator	28 Jan 1910 10:57 UT	23 Aquarius
Crossed ecliptic	19 Jan 1910 00:04 UT	02 Aquarius

Tropical Zodiac

Date	Sign
12 Jan 1910 - 18 Jan 1910	Capricorn
18 Jan1910 - 04 Feb 1910	Aquarius
04 Jan 1910 - 28 Feb 1910	Pisces

Constellations

Note: not to be confused with the 12 signs of the tropical zodiac.

Date	Constellation
12 Jan 1910 - 17 Jan 1910	Sagittarius
17 Jan 1910 - 20 Jan 1910	Capricorn
20 Jan 1910 - 01 Feb 1910	Aquarius
01 Feb 1910 - 28 Feb 1910	Pegasus

Planetary alignments

Date	Planet	Zodiac location
16 Jan 1910	Uranus	21 Capricorn
17 Jan 1910	Sun	26 Capricorn
21 Jan 1910	Mercury R	11 Aquarius
31 Jan 1910	Chiron	26 Aquarius
02 Feb 1910	Venus R	28 Aquarius

Note

In 1910 we had two great comets in the sky, the Great Comet of January 1910 and Halley's Comet in April. Their orbits are presented separately before analysing their astrological influence.

The Daylight Comet was one of the brightest ever witnessed, so bright it outshone Venus in the sky and was even visible during daylight hours, thus its name. I am estimating the final sighting being in late February, though there is no clear date for this.

Halley's Comet 1910 (1P/Halley)

Type of Comet: periodic

Orbital period: 72–80 years

Astronomical data

	Date	Position
First seen	10 Apr 1910	03 Aries
Perihelion	20 Apr 1910	01 Aries
Last day to be seen	11 June 1910	07 Virgo
Crossed celestial equator	06 June 1910 18:18 UT	05 Virgo
Crossed ecliptic		

Tropical Zodiac

Date	Sign
10 Apr 1910 - 16 May 1910	Aries
16 May 1910 - 19 May 1910	Taurus
19 May 1910 - 21 May 1910	Gemini
21 May 1910 - 24 May 1910	Cancer
24 May 1910 - 01 Jun 1910	Leo
01 Jun 1910 - 12 Jun 1910	Virgo

Constellations

Note: not to be confused with the 12 signs of the tropical zodiac.

Date	Constellation
10 Apr 1910 - 20 Apr 1910	Pisces
20 Apr 1910 - 29 Apr 1910	Pegasus
29 Apr 1910 - 16 May 1910	Pisces
16 May 1910 - 18 May 1910	Aries
18 May 1910 - 21 May 1910	Taurus
21 May 1910 - 23 May 1910	Gemini
23 May 1910 - 26 May 1910	Cancer
26 May 1910 - 29 May 1910	Hydra
29 May 1910 - 12 Jun 1910	Sextant

Fixed Stars

Date	Star	Zodiac location
22 May 1910	Alhena	Gemini
25 May 1910	Al Tarf	Cancer

Planetary alignments

Date	Planet	Zodiac location
16 May 1910	Saturn	29 Aries
19 May 1910	Sun	27 Taurus
19 May 1910	Mercury Rx	06 Taurus
21 May 1910	Pluto	25 Gemini
22 May 1910	Mars	12 Cancer
22 May 1910	Neptune	17 Cancer

Events

Two Great Comets appearing in the sky within three months of each other made 1910 an extraordinary year for stargazers, and inevitably it marked a time of great upheaval.

Unlike the Daylight Comet in January, Halley's Comet was already recognised as periodic and therefore its appearance had been well forecast, which helped it to become the first comet to be photographed and to provide spectroscopic data.[100]

It was also the source of mass panic when it was learned that its tail would actually pass through the Earth's atmosphere on 19 May at 25 Taurus, the degree of the fixed star Algol – the 'demon star' – where the Sun had been just three days earlier. Word spread that the tail contained poisonous gases that would kill all living beings on Earth, and opportunists profited from the panic by selling gas masks, anti-comet pills, helmets and even umbrellas, while householders stuffed paper around their doors and windows to seal them from the deadly threat.[101]

Blame for the rumour was laid at the door of French astronomer Camille Flammarion, who had discovered through spectroscopy that the deadly gas cyanogen lurked in the comet's tail. He denied newspaper reports that quoted him as saying the gas would "impregnate the atmosphere and possibly snuff out all life on the planet", but he is still blamed in some quarters for starting the panic.[102]

Looking at Flammarion's birth chart (Figure 8.6.1), we can see that the Great Daylight Comet first appeared beside his natal Saturn at 11 Capricorn, continuing past his natal Jupiter, North Node and Neptune, before entering Pisces and crossing his Venus-Sun conjunction (if my estimates for its final observation are correct). Meanwhile Halley's Comet, the source of the panic, passed by his Aries Mars and aligned with the Sun beside his Descendant at the tail end of Taurus.

Meanwhile the rumour of the comet's deadly fumes reached China, where they added fuel to the revolutionary fervour that threatened and

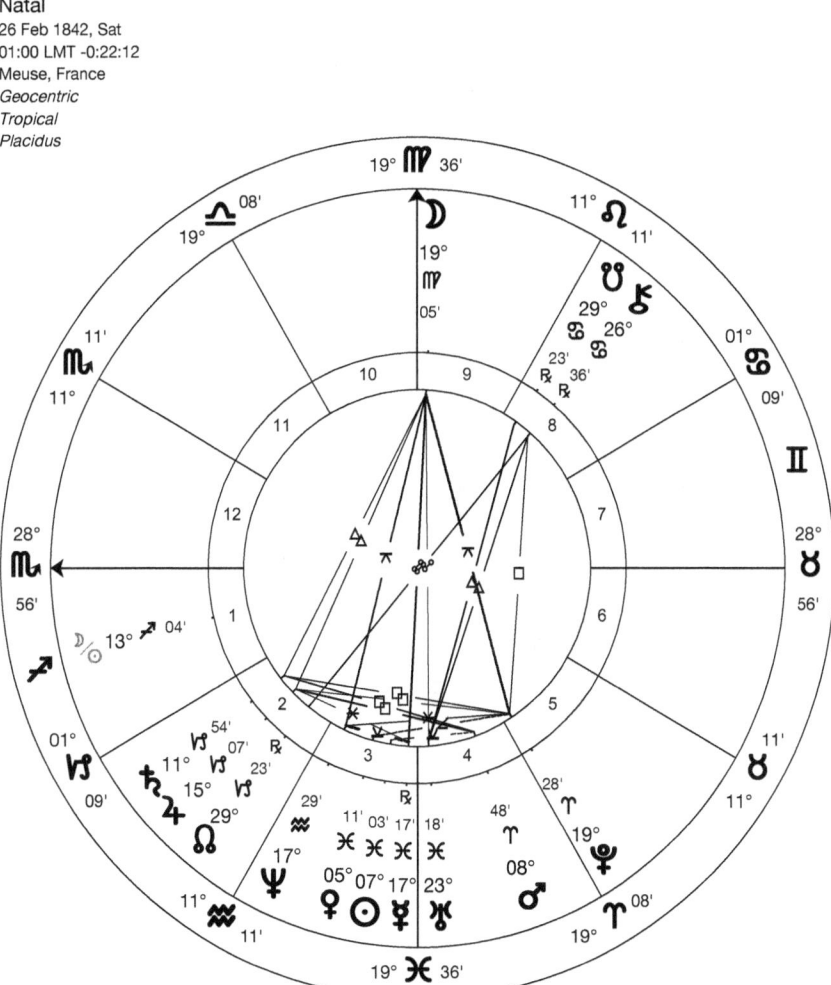

Figure 8.6.1

eventually brought down the fragile Qing dynasty, ending centuries of imperial rule with the abdication of the child emperor Puyi in 1912.[103] The comet was also seen as an omen of the Manchurian pneumonic plague that year, which killed 60,000 people and contributed to the general unrest.[104]

We can see from the birth chart of the Qing dynasty in 1636,[105] the Daylight Comet appeared at 10 Capricorn right on the empire's Saturn, while Halley's Comet crossed its Mercury, Chiron, Sun and Pluto in Taurus, as well as Venus in Cancer and Jupiter in Leo.

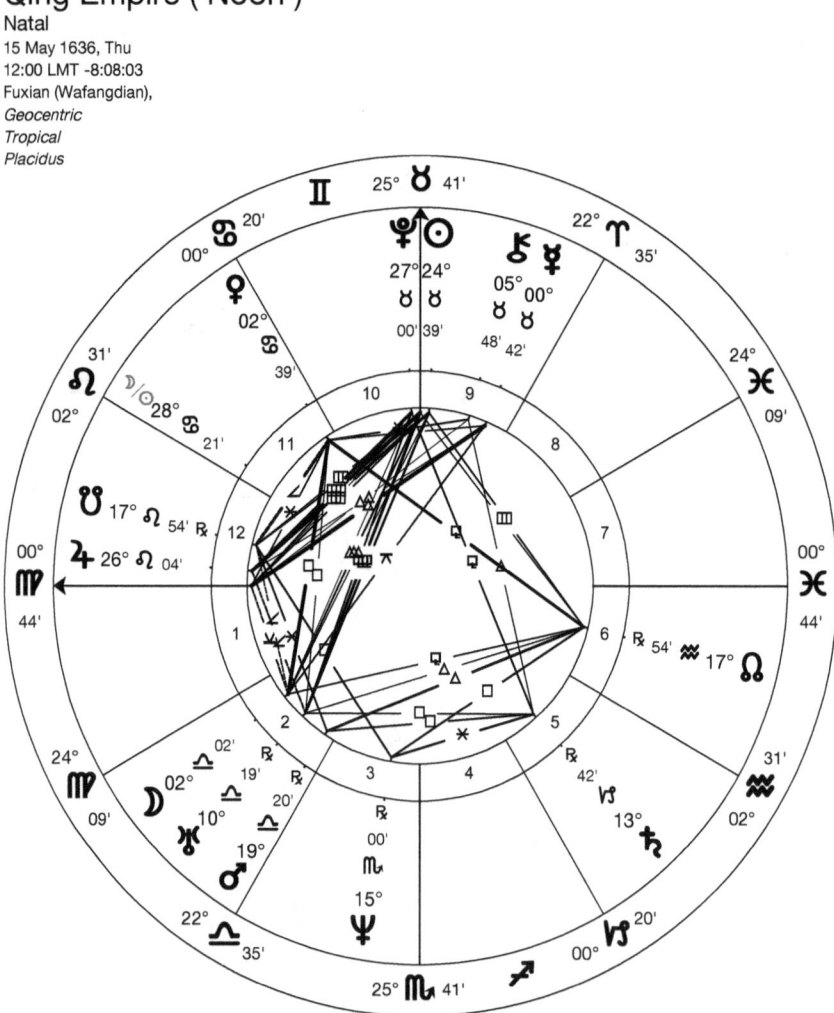

Figure 8.6.2

Looking at child emperor Puyi's natal chart for 7 February 1906,[106] we can see the Daytime Comet aligned with his natal Chiron, Mercury, Venus, Sun and South Node in Aquarius and his Saturn in Pisces. A few months later Halley's Comet conjoined Puyi's natal Mars when it was first observed and on the day the tail passed through the Earth's atmosphere at Algol's degree of 25 Taurus, it conjoined his Jupiter, before passing his Pluto, Neptune and Moon. (Figure 8.6.3).

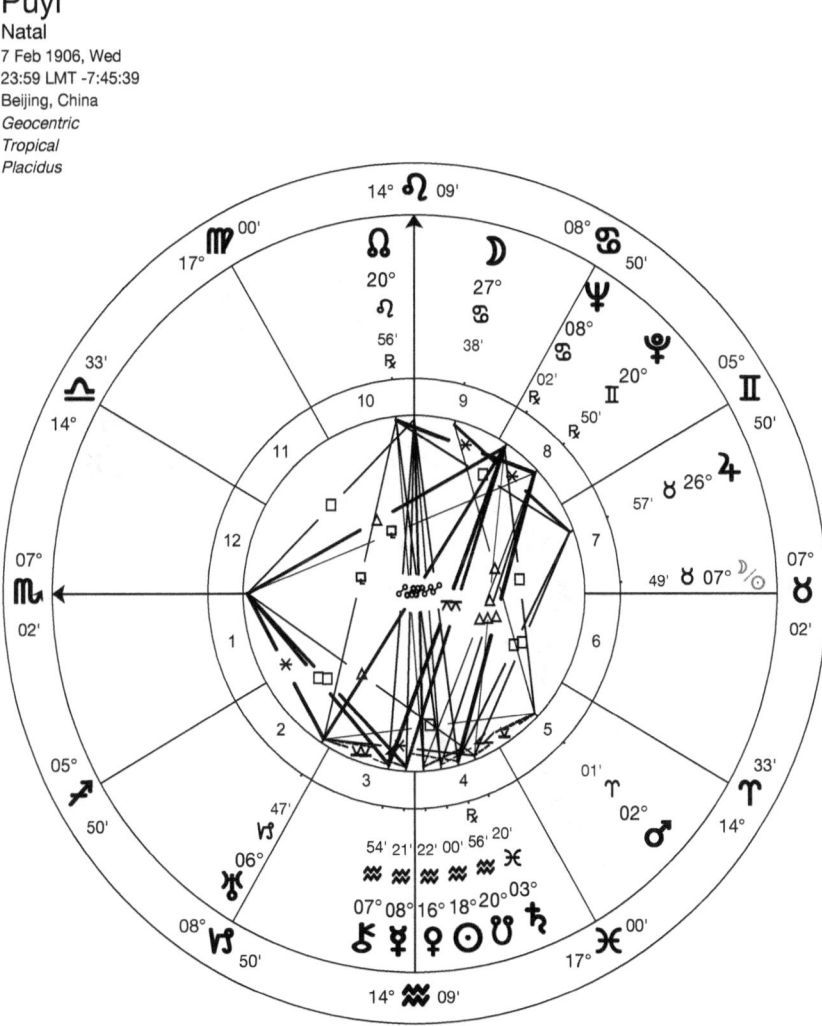

Figure 8.6.3

China was not the only empire to lose its monarch during this fateful period. King Edward VII, whose titles included Emperor of India, died on 6 May 1910 after a lengthy battle with bronchitis.[107] His natal North Node at 3 Aquarius marked the degree where the Daylight Comet crossed the ecliptic, while his son King George V who succeeded him had his Ascendant at 2 Aries just one degree from where Halley's Comet was first spotted. (Figure 8.6.4).

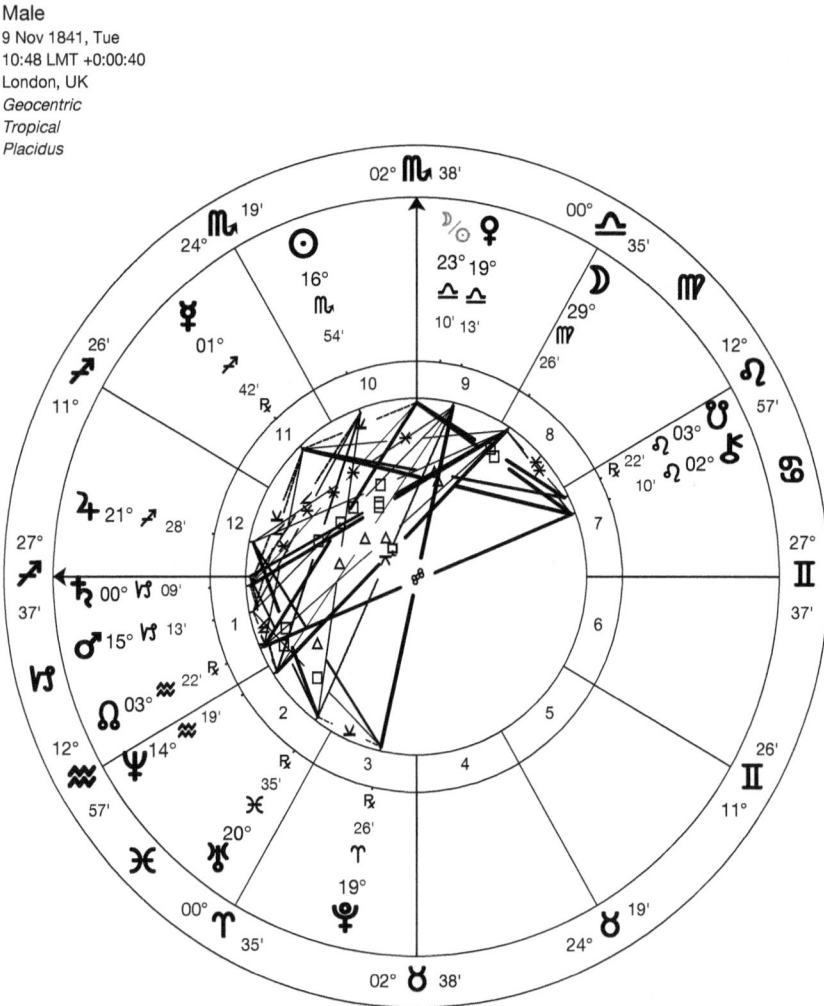

Figure 8.6.4

Other leaders who died during the year of the two Great Comets included:

- Egyptian prime minister Boutros Ghali (assassinated on 20 February);
- Panama president José Domingo de Obaldia (1 March); and
- King Chulalongkorn the Great of Siam, now Thailand (23 October). [108]

The famous American author Mark Twain also died that year. He was born two weeks after Halley's Comet's closest approach in 1835, and when he heard it was due to return, he announced: "It will be the greatest disappointment of my life if I don't go out with Halley's Comet. The Almighty has said, no doubt: 'Now here are these two unaccountable freaks; they came in together, they must go out together'."[109] There is little to connect the two 1910 comets to Twain's natal chart, apart from the Daylight Comet crossing the ecliptic by his natal Neptune at 1 Aquarius, and Halley's Comet swinging past his Pluto-Moon conjunction in Aries shortly after it was first seen.

We can observe astrological connections with some of the dramatic events of the period.

The Daylight Comet first appeared in the earth sign of Capricorn, which may correlate with the large number of coalmine explosions, including Colorado, Kentucky, Pennsylvania, Mexico and Lancashire, as well as a massive avalanche in Washington state that swept away two trains and killed 96 people, followed later by the worst snowslide accident in Canadian history.[110]

The air sign of Aquarius where the comet travelled next is often associated with aviation, technology and protest movements. Perhaps we can see a connection here with:

- the first commercial flight of the Zeppelin airship in Germany;
- Raymonde de Laroche becoming the first woman to receive a pilot's licence;
- Rolls-Royce co-founder Charles Rolls performing the first non-stop flight across the English Channel and back;
- the first take off from a ship taking place;
- cargo flights being introduced, with air mail; and

- the first ever aerial bombing mission arriving the following year.[111]

There were uprisings in the Ivory Coast, Albania, Morocco, Portugal, Brazil and the long and violent Mexican Revolution was launched; 20,000 women went on strike for better pay, shorter hours and union recognition at textile factories in New York; there were riots after African-American boxer Jack Johnson defeated his white-American opponent James J. Jeffries; and 300 suffragettes clashed with police outside the UK Parliament. The following year saw Britain's first ever national rail strike that led to riots in Wales where six people died. [112]

The wireless telegraph had become common by the late 19th century, but in 1910 it converted from mechanical to electrical transmission and became the standard mode of communication at sea. That year it played a role in the arrest of the infamous Dr Crippen, who became the first person to be captured using wireless telegraphy. Crippen, an American doctor living in London, had fled the country with his mistress after being questioned by the police about the disappearance of his wife Cora. However, when they boarded a transatlantic ship he was recognised by the captain, who sent a telegraph to the police that led to his arrest when they arrived in Canada. He was brought back to England where he was convicted of murder and executed after a four day trial, using evidence that a body found under the cellar floor in their house was that of his wife. It is interesting to note that in the year 2007 when Comet McNaught comet passed by, US forensic scientists ran DNA tests that proved the body had in fact been that of a male.[113] We may also note that Marconi sent the first wireless telegraph signals across the Atlantic in the year of the Great Comet of 1901.

Other events of the period I would associate with the two comets include: [114]

- The world's tallest building, the 700 foot Metropolitan Life Insurance Company Tower, was completed in New York City;
- Neon lights were demonstrated for the first time at the Paris Motor Show;
- The first Girl Guide troops were registered in the UK;
- Slavery was abolished in China after thousands of years;

- The first horror movie, *Frankenstein*, was released;
- Italy opened the first clinic to treat occupational diseases in Milan;
- Scott and Amundsen started their race to the South Pole, with Amundsen getting there first in 1911; and
- The first infrared photographs were published by Professor Robert Williams Wood in the Royal Photographic Society's journal.

Comet Ikeya–Seki (C/1965 S1)

Type of Comet: non periodic, Kreutz sungrazer

Astronomical data

	Date	Position
Discovery	18 Sept 1965	15 Leo
First seen	20 Oct 1965	22 Libra
Perihelion	21 Oct 1965 21:18 UT	27 Libra
Last seen	10 Dec 1965	14 Virgo
Crossed celestial equator		
Crossed ecliptic	21 Oct 1965 03:28 UT	27 Libra

Tropical Zodiac

Date	Sign
20 Oct 1965 - 21 Nov 1965	Libra
21 Nov 1965 - 10 Dec 1965	Virgo

Constellations

Note: not to be confused with the 12 signs of the tropical zodiac.

Date	Constellation
20 Oct 1965 - 25 Nov 1965	Virgo
25 Oct 1965 - 07 Nov 1965	Corvus
07 Nov 1965 - 11 Nov 1965	Crater
11 Nov 1965 - 23 Nov 1965	Hydra
23 Nov 1965 - 10 Dec 1965	Antlia (air pump)

Planetary alignments

Date	Planet	Zodiac location
21 Oct 1965	Sun	27 Libra
05 Dec 1965	Uranus	19 Virgo
06 Dec 1965	Pluto	18 Virgo

Events

The year Comet Ikeya-Seki appeared close to the Earth coincided with one of the most significant planetary aspects of the past century, the Uranus-Pluto conjunction in Virgo opposed by Saturn and Chiron in Pisces, which brought with it immense cultural shifts in society worldwide. The comet came within five degrees of the Uranus-Pluto conjunction on the last day it could be seen in the sky, adding extreme energy to the configuration.

At the time the US was caught up in the Vietnam War, sparking huge anti-war protests, the Cold War between NATO and the Soviet Union was at its height, and China was experiencing the tragedy of the Cultural Revolution.

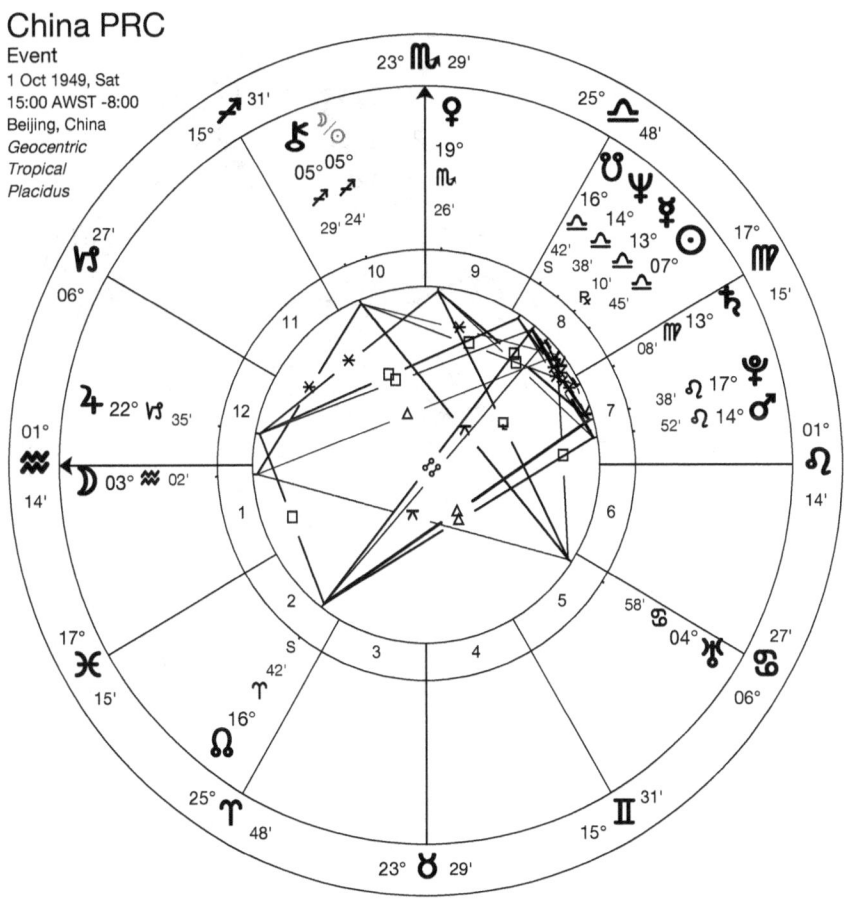

Figure 8.7.1

Launched by Chairman Mao Zedong in 1966 to protect the socialist state he had created to counter western influence, the revolution lasted 10 years and led to the death of millions of Chinese people, while disrupting the lives of millions more. It represented an inversion of the principles of justice and harmony represented by the sign of Libra, where the comet first appeared.

The comet's retrograde passage through the signs of Libra and Virgo saw it cross the South Node, Neptune, Mercury and the Sun in the 1949 chart of the People's Republic of China (Figure 8.7.1), symbolising loss, madness, the students who led the revolution and the country's leadership respectively. As the comet departed and transited that revolutionary

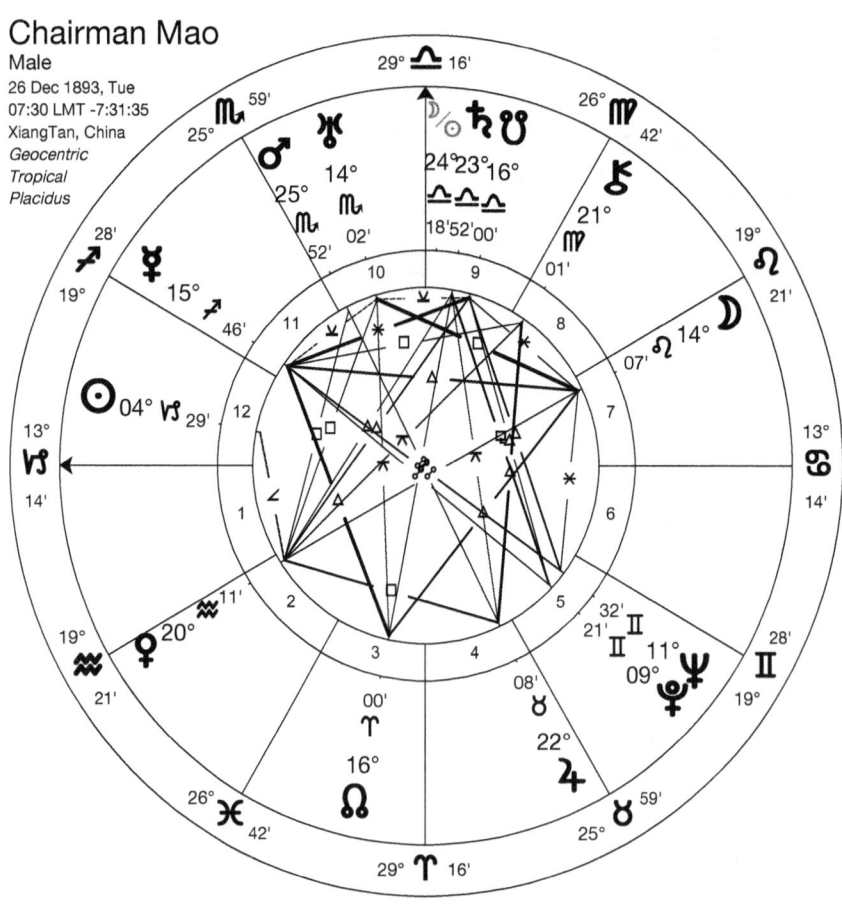

Figure 8.7.2

conjunction of Uranus-Pluto, it opposed Saturn, the planet of discipline, structure and repression, and Chiron, representing the wound.

Its discovery at 15 Leo aligns with Mao's Moon and China's Mars-Pluto conjunction, which jointly rule the country's Scorpio MC. Its final appearance at 14 Virgo aligns with China's Saturn.

The most potent degree in the comet's passage must be 27 Libra where it reached perihelion while crossing the ecliptic and aligning with the Sun, just two degrees from Mao's MC. We may note the Uranus-Pluto conjunction was very close to his Chiron at 21 Virgo as well. (Figure 8.7.2)

Comet Ikeya-Seki through the Elements

Between October and December while it was visible to the naked eye, this bright comet only travelled through the two signs Libra (air) and Virgo (earth).

As expected, the comet triggered a series of extreme air-related events, especially disasters:

- The two years[115, 116] following the comet's brief seven-week appearance in the sky during late 1966 saw an uncommon number of storms and plane crashes. Air India, All Nippon Airways, Canadian Pacific Air Lines, BOAC (later British Airways), British Midland Airways, United Airlines, Braniff International Airways, TABSO and Piedmont Airlines all experienced mass fatalities from crashes;

- The Topeka tornado, one of the worst experienced by the state of Kansas, killed 17 people and injured more than 500 in June 1966, causing major economic losses;[117]

- Hurricane Inez left more than 1,000 dead and tens of thousands homeless in Haiti and the Dominican Republic;

- In January 1967 the worst blizzard in Chicago's history dumped 23 inches of snow in 35 hours bringing the city to a standstill, followed by an outbreak of tornadoes on 21 April killing 33 people and injuring 500 in the city and surrounding Midwest;

- The same month three US astronauts died when their spacecraft caught fire in a launch simulation at a time when the space race with the Soviet Union was hotting up. The following April the first Soviet cosmonaut died re-entering Earth's orbit and another US astronaut died later that year;

- The US, Soviet Union and the UK signed the Outer Space Treaty banning weapons of mass destruction from space;
- In 1965 France launched a satellite to become the third country to enter outer space, the following February the Soviets landed the first unmanned spacecraft on the Moon soon to be followed by the US, and in March their space probe Venera 3 crashed into Venus; and
- The Beatles' song 'All You Need Is Love' was premiered on the first ever live international satellite TV broadcast watched by 400 million viewers.

Libra also symbolises issues of international, cultural and social relations and this period certainly saw plenty of upheaval in this field:

- China saw the beginning of the tragic Cultural Revolution (see above);
- The anti-war movement took shape in the US[118], along with the civil rights movement[119] and the hippie counterculture that led to major clashes with authority and the banning of the psychedelic drug LSD;
- An anti-communist purge in Indonesia saw at least 500,000 to 1 million people killed;
- Hong Kong experienced a series of violent riots against the British Hong Kong government in which 51 people died;[120]
- Moscow experienced the Soviet Union's first spontaneous political demonstration for civil rights called the 'Glasnost Meeting';
- There were coups in the Central African Republic, Nigeria, Argentine, Jamaica, Togo and Greece;
- White controlled Southern Rhodesia introduced martial law and declared independence from the British Empire in November 1965 in an effort to delay black majority rule;[121]

- President Charles de Gaulle took France out of NATO and closed all its bases in the country and NATO headquarters moved to Brussels;
- Civil rights lawyer Thurgood Marshall[122] became the first African-American justice of the US Supreme Court in August 1967 amidst fierce debate, while huge race riots broke out throughout the US;
- The same year the court declared all US state laws banning interracial marriage unconstitutional, while two days before the comet disappeared from view in December 1965 the UK signed the Race Relations Act and two weeks later the UN adopted the international convention against racial discrimination;
- In 1966 India and Pakistan signed a peace agreement, and Indonesia and Malaysia ended three years of hostility with a joint peace declaration;
- In 1967 Britain decided to join the European Economic Community, legalised abortion and decriminalised homosexuality; and
- The word 'fuck' was broadcast for the first time in a BBC discussion programme by theatre critic Kenneth Tynan, causing outrage (and no doubt some amusement) throughout the country.

If we relate the earth element to geological and economic matters, we can see some significant events during this period of upheaval.

- In December 1965 as the comet disappeared from sight, the United Nations established the World Food Programme as a permanent agency;[123]
- Earthquakes causing mass casualties occurred in China, Turkey and Venezuela in 1966 and 1967;
- On 10 July 1967 heavy rains and a landslide in Kobe and Kure, Hiroshima, Japan, killed at least 371 people;
- The Aberfan disaster in Wales killed 144 people, including 116 children, when a coalmine spoil tip collapsed;

- The United Kingdom announced moves to convert sterling to a decimal currency, and the Australia dollar was launched;[124]
- In December the UK froze Rhodesian reserves and imposed total economic sanctions against the white regime that had declared independence;
- The biggest oil spill in history so far occurred when the supertanker Torrey Canyon ran aground off Britain's south west coast; and
- The world's first cash machine was opened in the UK in 1967.

If we include the fire sign of Leo, where the comet was when it was first discovered by the astronomers Ikeya and Seki, we can note:

- A fire destroyed almost three quarters of the Philippines city of Iloilo;
- Severe bush fires raged across more than 2,500 square kilometres in southern Tasmania claiming 62 lives; and
- 323 people were killed or went missing and 150 were injured in Belgium's most devastating fire at the Innovation department store in Brussels.

One more fascinating correlation relates to the constellation Antlia, which the comet passed through as it disappeared from view. The Greek word *antlia* means 'pump', and December 1967 saw the world's first transplant of a human heart – which, of course, pumps blood around the body - performed by South African surgeon Christiaan Barnard.[125]

Halley's Comet of 1986

Astronomical data

	Date	Position
Discovery	Not known	
First seen	08 Nov 1985	14 Gemini
Perihelion	09 Feb 1986	15 Aquarius
Last seen	End April-early May	Virgo
Crossed celestial equator	24 Dec 1985	11 Pisces
Crossed ecliptic	09 Nov 1985 02 56 UT	11 Gemini

Next perihelion 28 Jul 2061

Tropical Zodiac

Date	Sign
01 Nov 1985 - 16 Nov 1985	Gemini
16 Nov 1985 - 27 Nov 1985	Taurus
27 Nov 1985 - 10 Dec 1985	Aries
10 Dec 1985 - 10 Jan 1986	Pisces
10 Jan 1986 - 11 Mar 1986	Aquarius
11 Mar 1986 - 02 Apr 1986	Capricorn
02 Apr 1986 - 10 Apr 1986	Sagittarius
10 Apr 1986 - 16 Apr 1986	Scorpio
16 Apr 1986 - 26 Apr 1986	Libra
26 Apr 1986 - 30 Apr 1986	Virgo

Constellations

Note: not to be confused with the 12 signs of the tropical zodiac.

Date	Constellation
01 Nov 1985 - 18 Nov 1985	Taurus
18 Nov 1985 - 27 Nov 1985	Aries
27 Nov 1985 - 20 Dec 1985	Pisces
20 Dec 1985 - 23 Feb 1986	Aquarius
23 Feb 1986 - 12 Mar 1986	Capricorn

12 Mar 1986 - 30 Mar 1986	Sagittarius
30 Mar 1986 - 02 Apr 1986	Corona Australis
02 Apr 1986 - 06 Apr 1986	Scorpio
06 Apr 1986 - 07 Apr 1986	Ara
07 Apr 1986 - 09 Apr 1986	Norma
09 Apr 1986 - 12 Apr 1986	Lupus
12 Apr 1986 - 18 Apr 1986	Centaurus
18 Apr 1986 - 24 Apr 1986	Hydra
24 Apr 1986 - 30 Apr 1986	Crater

Planetary alignments

Date	Planet	Location
08 Nov 1985	Chiron	13 Gemini
23 Jan 1986	Jupiter	23 Aquarius
03 Feb 1986	Venus	17 Aquarius
04 Feb 1986	Mercury	17 Aquarius
05 Feb 1986	Sun	16 Aquarius
31 Mar 1986	Neptune	5 Capricorn
01 Apr 1986	Mars	1 Capricorn
05 Apr 1986	Uranus	22 Sagittarius
08 Apr 1986	Saturn	9 Sagittarius
14 Apr 1986	Pluto	6 Scorpio

This was the most recent appearance of Halley's Comet. It will not reach our part of the solar system again until 2061 in its three quarters of a century orbit. This comet has been observed and recorded for more than 2,000 years, but this was the first time it could be examined in detail from space. The European Space Agency spacecraft *Giotto* found the comet was composed of a mixture of volatile ices – such as water, carbon dioxide and ammonia – and dust, which makes up most of the comet's surface. It is therefore known by astronomers as a "dirty snowball". The comet was also examined by the Japanese space probe *Suisei*.

In 1986 the comet disappointed Earth-bound astronomers and amateurs alike as it lacked a tail and merely appeared as a small bright dot in the sky. It was visible to the naked eye from November 1985 to April 1986.

Events

Soviet Union

History tells us that the events that led up to the dissolution of the Soviet Union in 1991 began during the period influenced by Halley's Comet appearing in the sky, with the appointment of new leader Mikhail Gorbachev in March 1985, and the 'glasnost' and 'perestroika' reforms he announced in February 1986 designed to create more freedom, end corruption and modernise the Soviet economy.

During this period the Soviet Union started pulling out of its occupation of Afghanistan and allowed more open contact with the west, holding meetings with US president Ronald Reagan and UK prime

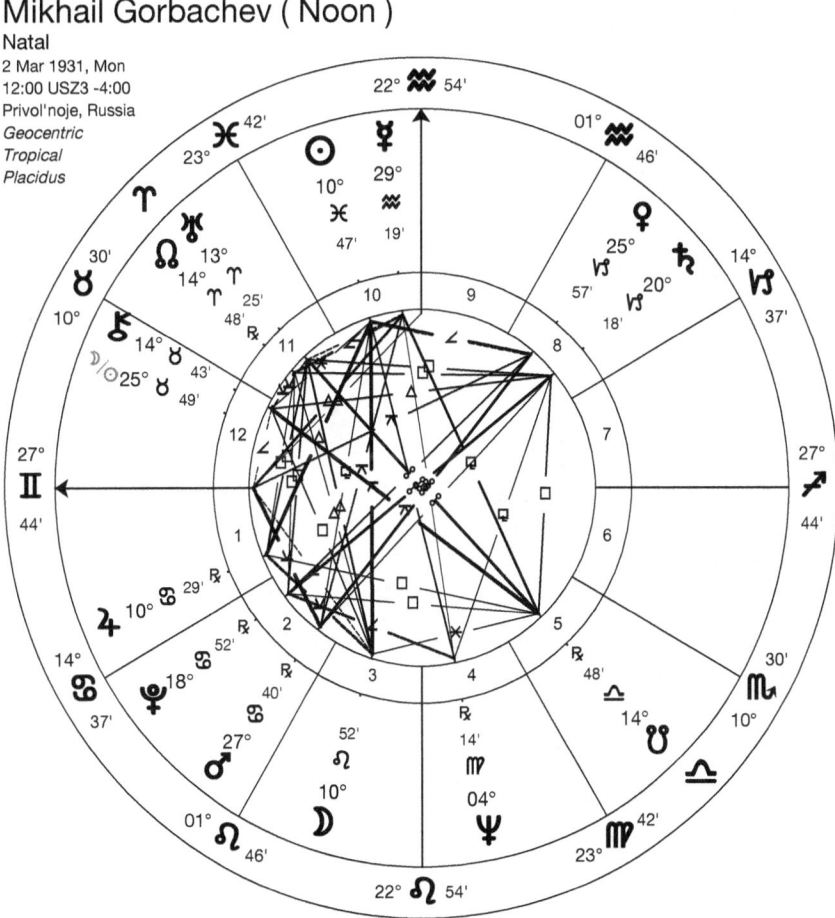

Figure 8.8.1

minister Margaret Thatcher (who even appeared on Soviet TV), and signing an arms control treaty with the US.

Loosening the controls led to the first shoots of resistance appearing in the Baltic States with the "Singing Revolution" and the Phosphorite War in Estonia and with mass protests in Romania, which snowballed over ensuing years. An anti-government riot in the state of Kazakh, now Kazakhstan, left 120 dead. During this time Reagan challenged Gorbachev to "tear down" the Berlin Wall, which happened at the hands of the East German people themselves in 1989 to become the event that symbolised the end of Communist rule in Eastern Europe.

Neptune's recent entry into Capricorn clearly had an influence on the loosening of political control, as did the first Jupiter-Saturn square after their

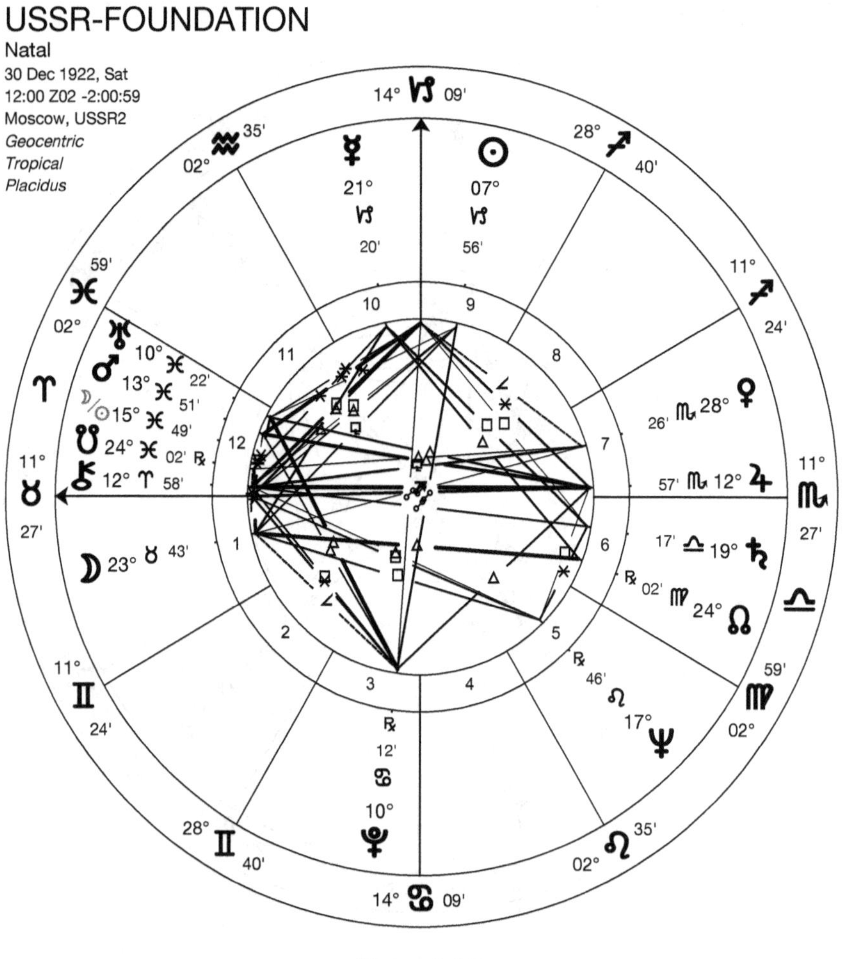

USSR-FOUNDATION
Natal
30 Dec 1922, Sat
12:00 Z02 -2:00:59
Moscow, USSR2
Geocentric
Tropical
Placidus

Figure 8.8.2

conjunction in Libra in 1982 – the first in an air sign for hundreds of years.

In terms of Halley's Comet's influence, it crossed the celestial equator at 11 Pisces within a degree of Gorbechev's Sun at 10 Pisces, the same degree of Uranus in the 1922 founding chart of the Soviet Union and close to that chart's Mars at 13 Pisces. (Figures 8.8.1 and 8.8.2)

Chernobyl

The explosion at the Chernobyl nuclear power plant in the Soviet Union, in what is now Ukraine, on 26 April 1986 remains human history's worst disaster in terms of cost alone. An estimated $700 billion has been spent on repairing the damage it caused. The response involved more than 500,000 personnel.

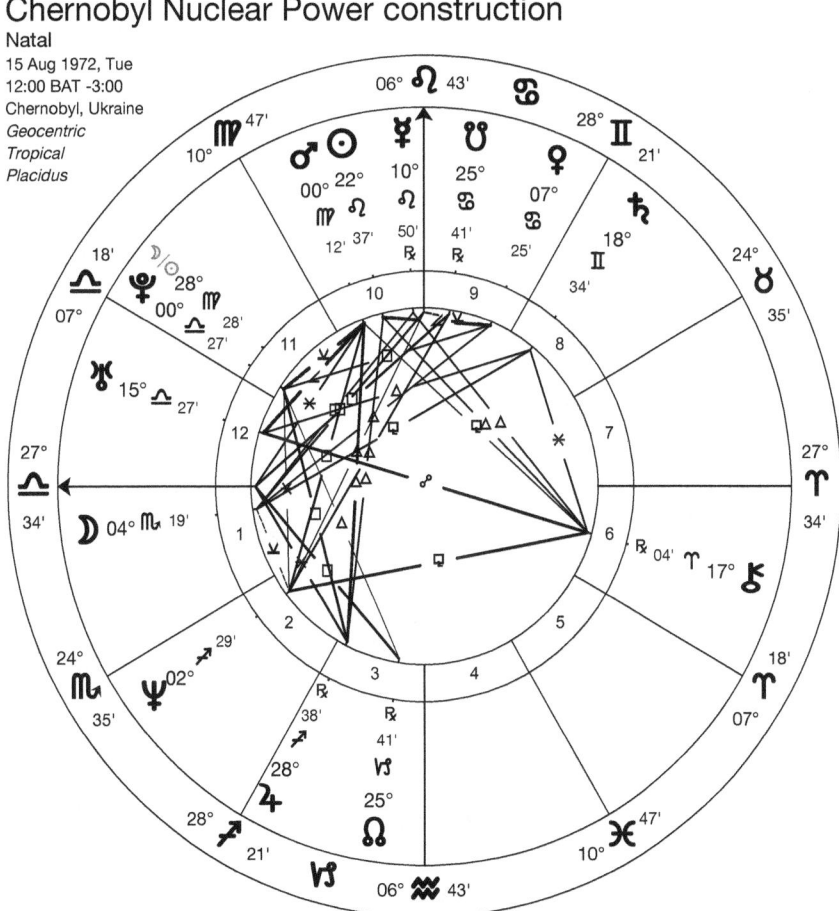

Figure 8.8.3

98   *Comets in Astrology*

The explosion occurred when a safety test went wrong, components ruptured, coolant leaked and the resulting explosions caused a meltdown that destroyed a reactor and started a fire that spread radioactive contaminants across the Soviet Union, Europe and further afield.

More than 100,000 people were evacuated, two engineers were killed, two were severely burned, 237 workers were hospitalised, 28 of whom died within three months. Estimates for the number of deaths from related sickness range from the tens of thousands to a million.

Former Soviet president Gorbachev stated later that Chernobyl was probably the real cause of the collapse of the Soviet Union five years later.

The explosion occurred just as Halley's Comet was disappearing from view having just passed through the constellation of Hydra, symbolising

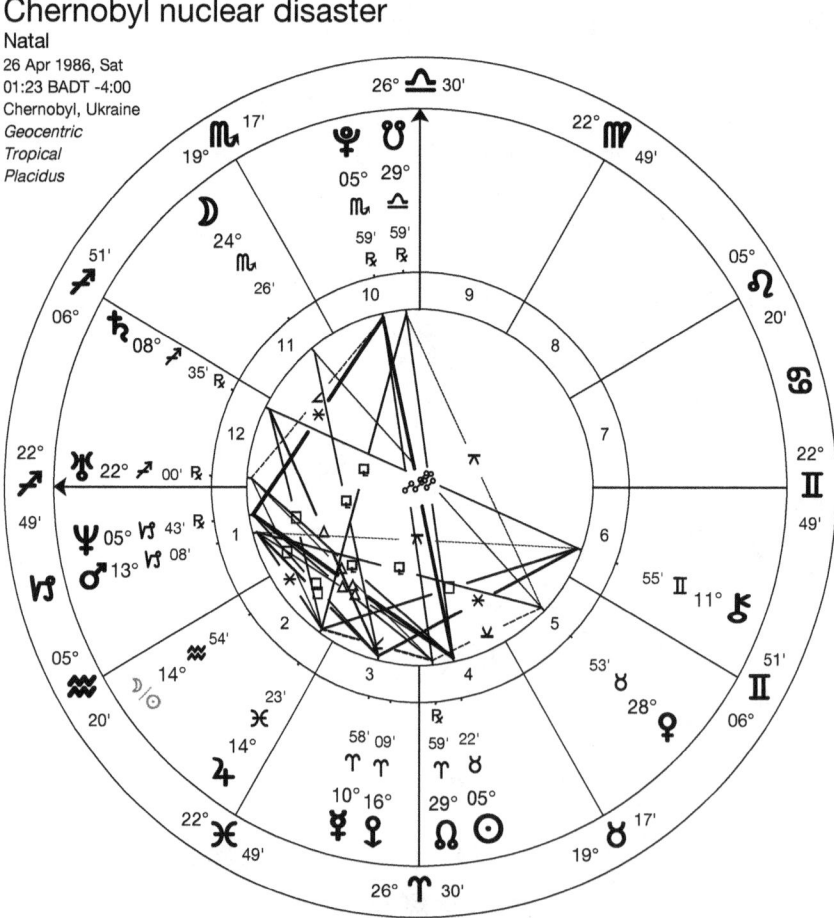

Figure 8.8.4

the many-headed monster. The previous month the comet had aligned with Mars, Saturn, Uranus, Neptune and Pluto. On the day of the disaster Chiron was transiting at 11 Gemini where the comet crossed the ecliptic. That same day the comet itself was on the World Axis at 0 Libra, the same degree Pluto was positioned when construction began on the power plant in 1972. (Figures 8.8.3 and 8.8.4)

Space Shuttle 'Challenger'

Three months before the Chernobyl disaster, viewers across the globe (including my childhood self) were transfixed and horrified to watch on live television the dramatic explosion of the US Space Shuttle *Challenger* just 73 seconds after take-off. All seven crew, including 37 year old teacher

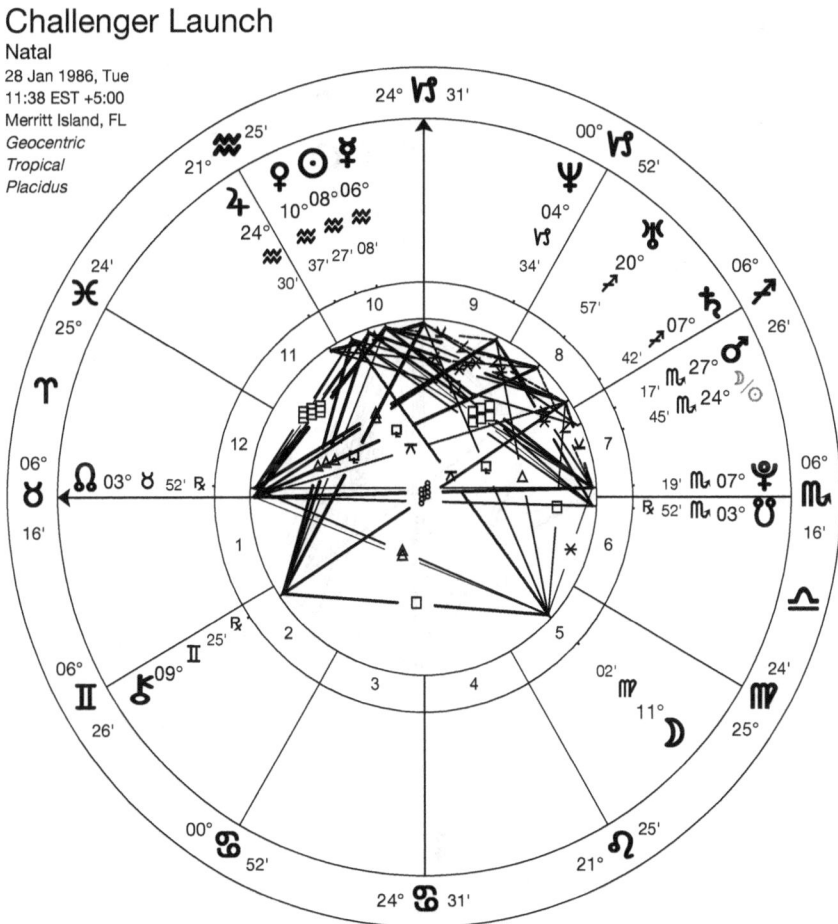

Figure 8.8.5

Christa McAuliffe, died in the accident caused by faulty seals in the spacecraft's right rocket booster.

In the chart for that fateful launch we see a Mercury-Sun-Moon stellium in Aquarius squaring Pluto on the Descendant at 7 Scorpio, while Jupiter at 24 Aquarius squares Mars in Scorpio – both signatures associated with explosions and air. Halley's Comet aligned with Jupiter at 23 Aquarius and with Pluto at 6 Scorpio, activating both 'challenging' aspects in the chart. (Figure 8.8.5)

The comet crossed the ecliptic at 11 Gemini where it aligned with transiting Chiron and opposed the Jupiter-Uranus conjunction in Sagittarius that can be found in the chart of the spacecraft's first mission in April 1983. (Figure 8.8.6)

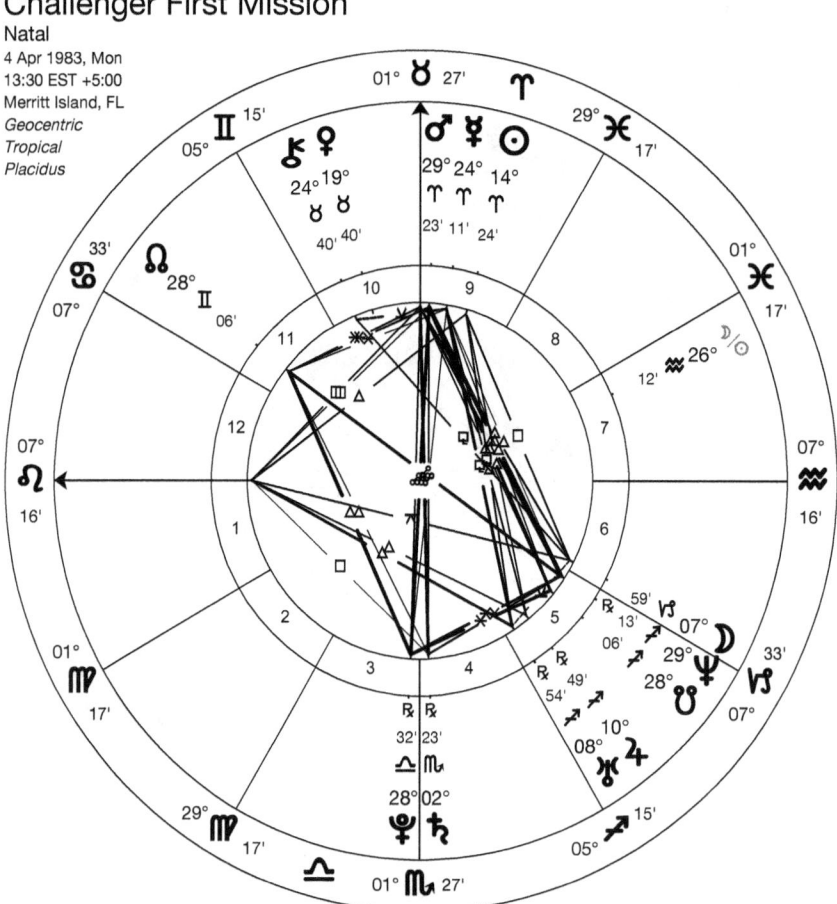

Figure 8.8.6

Halley's Comet through the Elements

During its six month visit between November 1985 and April/May 1986, Halley's Comet passed though nine of the 12 signs of the zodiac between Gemini and Virgo, taking in all of the four elements.

Fire

The same week that the comet was first seen in November 1985, Columbia experienced two great tragedies: the Nevado del Ruiz volcano erupted killing 23,000 people, including almost the entire population of the town of Armero, in the second deadliest volcanic eruption of the 20th century; a few days before around 100 people, including half of the country's top judges, were killed in a violent guerilla attack on Columbia's Supreme Court. We may note that the chart for the founding of Columbia has Jupiter at 15 Aquarius, where the comet reached perihelion this time around.

This entire three year period was marked by an astonishing level of acts of violence, mostly politically motivated, in almost every corner of the world:

- A vast number of kidnappings, plane and ship hijackings, bombings and murders took place in the Middle East and around the world as Islamic extremists, many funded by Libya, sought retribution against US involvement in the region, resulting in the deaths of hundreds of people;
- The period ended with the launch of the first "intifada" which lasted for a further six years and led to more than 1,000 Palestinians, 100 Israeli civilians and 60 Israeli soldiers being killed;
- Nepal saw a failed attempt to start a communist revolution that left at least eight people dead;
- The IRA in Northern Ireland continued its long-running bombing campaign in the UK;
- Sikh terrorists blew up an Air India flight killing more than 300 people;
- South Africa declared a state of emergency as violence spread throughout the country in opposition to the apartheid regime;

- Hundreds of political prisoners in Peru were killed during riots;
- Sweden's premier Olaf Palme was murdered, as was Seychelles opposition leader Gérard Hoarou, American gorilla conservationist Dian Fossey, Palestinian cartoonist Naji al-Ali and five died in the failed attempt to assassinate former Chilean dictator Augusto Pinochet;
- Deadly civil wars began in South Yemen and Somalia;
- France joined the war with Chad against Libya;
- 8,000 died in riots in Egypt by military conscripts;
- Many people in the US, Australia and the UK were killed in shooting sprees and bombings by lone attackers, including the US anti-technology serial killer 'Unabomber';
- China's military killed 24 Vietnamese refugees at Lieyu Island;
- The Tamil Tigers commenced a bombing campaign in Sri Lanka that killed over a hundred people in one bus blast alone;
- The conflict between Iraq and Iran continued with Iraq dropping mustard gas on Iran and attacking a US frigate with missiles;
- 75 Indian Muslims were killed by police during riots in Kashmir;
- Basque separatists killed 21 people in a Barcelona bomb attack;
- Saudi security forces killed 400 Iranian pilgrims in Mecca;
- North Korea bombed a South Korean plane, killing 115 people; and
- French government secret agents sank the Greenpeace campaign vessel *Rainbow Warrior* to halt their campaign opposing nuclear tests in the Pacific, drowning a Portuguese photographer.

Meanwhile there were three major fires in the UK at Bradford City football stadium (56 dead, 265 injured), Manchester airport (55 deaths) and King's Cross underground station (31 deaths, 100 injured). A riot at the European Cup final at Belgium's Heysel stadium between Liverpool and Juventus resulted in 39 deaths and 600 injuries. In Puerto Rico an arson attack on a hotel killed almost 100 and injured 140 people, in the second deadliest hotel fire in US history.

There were coups in Burundi, Burkina Faso, Fiji and the People's Power Revolution in the Philippines overthrew the dictatorship of Ferdinand Marcos after 14 years in power.

Earth

The Earth itself rumbled significantly during this period with great loss of life. At least 5,000 people died and 30,000 were injured when a violent, "maximum intensity" earthquake struck off Mexico bringing down buildings in Mexico City causing damage estimated at $5 billion.

An earthquake in Chile left 177 dead, 2,575 injured, destroyed almost 150,000 homes, and left a million people homeless, while 1,500 people were killed in an earthquake in El Salvador. There were also major earthquakes in Greece and Bulgaria with fewer deaths and injuries.

Two rare limnic eruptions under two lakes in Cameroon asphyxiated almost 2,000 people and 3,000 livestock with carbon dioxide.

A hotel in Singapore collapsed leaving 33 dead and 17 injured, a bridge in the US collapsed killing 10 and in Italy a dam collapsed resulting in 248 deaths, 63 buildings destroyed and eight bridges demolished. Another dam in Argentine also collapsed and flooded a largely uninhabited village popular with tourists.

Financially the stock markets saw the Dow Jones Industrial Average peak at its highest level after a five year long bull market more than tripled its value. But fears of overvaluation combined with a decline in the value of the dollar led three months later to 1987's Black Monday, which saw stocks worldwide lose $1.7 trillion in value, a wake-up call that sparked fears of a global great depression.

The same period saw Communist China dip its toe into the capitalist market for the first time by creating its first share-holding commercial bank, the China Merchants Bank, which is now the world's 26th largest bank with more than 500 branches in China alone, as well as branches in New York and London.

In the UK three ground-breaking construction projects were underway with the Channel Tunnel, London's Docklands Railway and the first ring road around the capital, the M25. The Thatcher government also privatised

British Airways, while Egyptian businessman Mohamed al-Fayed purchased luxury department store Harrods.

Air

This was an unusually dangerous time for air transport with 19 accidents and around 2,150 lives lost. There were crashes in Spain, China, the Soviet Union, the US, Canada, Mexico, Mozambique, Scotland, Poland, the Philippines, Mauritius, Peru and Japan, where 520 people died in the worst single aircraft accident in aviation history. In the same period Emirates Airways was launched.

There were also three major train crashes, one in Canada that killed 23 and injured 71 people, one in Maryland where 16 people died and one in Indonesia with over 100 lives lost.

Extreme weather saw 10,000 people in Bangladesh die during an intense tropical storm, and 92 people die in a hailstorm with hailstones weighing up to 1kg. A burst of 44 tornados in the US killed 90 people; 12 inches of rain fell in a single day on Sydney, Australia; a 21 day tropical cyclone over the South China Sea killed 490 people in the Philippines, Taiwan, China and Vietnam; a tornado in Canada killed 21 people and left hundreds injured and homeless; 812 people died when Typhoon Nina brought 165mph winds to the Philippines; and 22 died in Britain and France when winds reached 120mph in what was dubbed the Great Storm of 1987.

This was also the time when concern about human impact on the planet's atmosphere increased significantly as a result of the British Antarctic Survey discovering the hole in the ozone layer.

In the world of digital technology we saw some early stutters of the coming internet revolution. The internet domain name system was created, Microsoft released Windows 1.0, the first email management software came out, Apple introduced its Hypercard, a precursor of the World Wide Web, and the first commercial 3D printer was sold.

In the world of international relations, western Europe moved ever closer to integration with Spain and Portugal joining the European Commission (now the European Union), the Single European Act created a single internal market and the Schengen Agreement abolished border controls between member countries.

In contrast the Iran-Contra affair exposed US government double dealing, with senior officials breaching their own arms embargo to sell weapons to Iran, using the proceeds to fund rebels fighting to depose the left wing government in Nicaragua.

In space, as well as the *Challenger* disaster, the period saw Japan launch the first deep space probe not run by the US and USSR, the first Muslim entered space travel, the Space Shuttle *Atlantis* had its maiden flight, the Soviet Union launched it space station *Mir* and a supernova became visible to the naked eye for the first time since 1604.

A famine in Ethiopia that affected one fifth of its 40 million people inspired the Live Aid concerts in London, Philadelphia, the Soviet Union, Canada, Japan, Yugoslavia, Austria, Australia and West Germany. Thanks to one of the largest ever satellite link ups and television broadcasts of all time, it attracted an estimated live audience of 1.9 billion people in 150 nations, nearly 40 per cent of the world's population at the time. The event raised more than £50 million for famine relief.

On 16/17 August 1987 the world's first synchronised global peace meditation took place. Organisers of the Harmonic Convergence said the date was selected according to Maya and Aztec cosmology and to take advantage of a wide Grand Trine between all the planets and the Sun, (but not the Moon). They hoped it would help to shift the world from an age of conflict to one of cooperation.

And in modern culture we saw the launch of *Neighbours*, one of the world's longest running soap operas; Japanese cartoon production firm Studio Ghibli; US cartoon creator Pixar; the first Simpsons cartoons; the Fox Broadcasting Company; Japanese video game developer Nintendo and its central character *Super Mario*; the longest running Broadway show *Phantom of the Opera*; and the world's largest professional wrestling promotion *WrestleMania* came into being at Madison Square Gardens in New York.

Water

This was also a terrible time for shipping disasters, including the deadliest peacetime maritime disaster in history when almost 4,400 lives were lost after an overcrowded passenger ferry collided with an oil tanker in the Philippines.

Elsewhere more than 500 people died when a ferry capsized off Bangladesh, nearly 400 died when two Soviet ships collided in the Black Sea, and almost 200 passengers and crew died when the British cross channel ferry *Herald of Free Enterprise* capsized off Zeebrugge in Belgium.

In the US state of Georgia, 2,000 tons of oil was spilled into the Savannah River, a Soviet tanker spilled up to 650 tons of oil off Finland and a chemical spill caused by a fire at a storehouse run by Sandoz in Switzerland caused mass wildlife mortality in the Rhine.

After decades of searching and many attempts, the wreck of the *Titanic* was finally discovered off the coast of Newfoundland, 12,000 feet below the surface. It has since been repeatedly visited by explorers, scientists, filmmakers, tourists and salvagers with thousands of artefacts recovered and put on display.

Comet Hale–Bopp (C/1995 O1)

Type of Comet: long periodic

Orbital period : 2,456-2,533 years

Astronomical data

	Date	Position
Discovery	23 Jul 1995	29 Cancer
First seen	20 May 1996	23 Capricorn
Last perihelion	1 Apr 1997	11 Aries
Last seen	9 Dec 1997	08 Virgo
Crossed celestial equator	26 Jun 1997 14:42UT	09 Cancer
Crossed ecliptic	6 May 1997 13:35 UT	15 Gemini

Tropical Zodiac

Date	Sign
20 May 1996 - 09 Aug 1996	Capricorn
09 Aug 1996 - 27 Nov 1996	Sagittarius
27 Nov 1996 - 31 Jan 1997	Capricorn
31 Jan 1997 - 25 Feb 1997	Aquarius
25 Feb 1997 - 11 Mar 1997	Pisces
11 Mar 1997 - 25 Mar 1997	Aries
25 Mar 1997 - 15 Apr 1997	Taurus
15 Apr 1997 - 04 Jun 1997	Gemini
04 Jun 1997 - 16 Aug 1997	Cancer
16 Aug 1997 - 01 Nov 1997	Leo
01 Nov 1997 - 09 Dec 1997	Virgo

Constellations

Note: not to be confused with the 12 signs of the tropical zodiac.

Date	Constellation
01 May 1996 - 27 Jun 1996	Sagittarius
27 Jun 1996 - 23 Jul 1996	Scutum
23 Jul 1996 - 05 Aug 1996	Serpens
05 Aug 1996 - 23 Nov 1996	Ophiuchus
23 Nov 1996 - 05 Dec 1996	Serpens

05 Dec 1996 - 16 Dec 1996	Ophiuchus
16 Dec 1996 - 07 Jan 1997	Serpens
07 Jan 1997 - 03 Feb 1997	Aquila
03 Feb 1997 - 12 Feb 1997	Sagitta
12 Feb 1997 - 23 Feb 1997	Vulpecula
23 Feb 1997 - 06 Mar 1997	Cygnus
06 Mar 1997- 14 Mar 1997	Lacerta
14 Mar 1997- 07 Apr 1997	Andromeda
07 Apr 1997- 23 Apr 1997	Perseus
23 Apr 1997- 19 May 1997	Taurus
19 May 1997 - 17 Jun 1997	Orion
17 Jun 1997- 27 Jul 1997	Monoceros
27 Jul 1997 - 08 Aug 1997	Canis Major
08 Aug 1997 - 14 Oct 1997	Puppis
14 Oct 1997 - 03 Nov 1997	Vela
03 Nov 1997 - 09 Dec 1997	Carina

Fixed Stars

Date	Star
2 Apr 1997	Almach
11 Apr 1997	Algol
3 Jun 1997	Betelgeuse

Planetary alignments

Date	Planet	Zodiac location
02 Jul 1996	Jupiter	13 Capricorn
04 Dec 1996	Mercury	1 Capricorn
01 Jan 1997	Sun and Mercury (Rx)	11 Capricorn
28 Jan 1997	Neptune	27 Capricorn
04 Feb 1997	Jupiter	03 Aquarius
06 Feb 1997	Venus	04 Aquarius
06 Feb 1997	Uranus	05 Aquarius
04 Mar 1997	Sun	14 Pisces
15 Mar 1997	Saturn	8 Aries
04 Jun 1997	Venus	29 Gemini
28 Jun 1997	Mercury	10 Cancer
04 Jul 1997	Sun	12 Cancer

Comet Hale-Bopp was first seen with the naked eye on 20 May 1996 in Brisbane, Australia by Terry Lovejoy.[126] As it approached the Sun in December it became difficult to see, but in January 1997 it was clearly visible again, its brightness increasing until it reached magnitude -1.8 as it passed perihelion on 1 April.[127]

After perihelion the comet moved into the southern half of the sky so it could be seen from the Earth's southern hemisphere, gradually fading during the summer and autumn until it disappeared from view after 9 December that year.[128] Thus it was visible without instruments for a total of 569 days, or 18.5 months, more than twice the record set by the Great Comet of 1811 of nine months.[129] As a result of its long passage close to Earth, it passed through 10 of the 12 zodiac signs, 18 separate constellations and aligned with three fixed stars, while the only planets it did not align with are Mars and Pluto. Therefore it is no surprise that we can find a large number of comet-related events during the period Comet Hale-Bopp was traversing our heavens.

Events

This period was marked by an outstanding number of disasters, conflicts and historical events, many of which had long term consequences that have continued to be felt well into the 21st century. The number of air disasters was quite unprecedented and several important political shifts took place in all corners of the world, including the bloodiest war since World War Two in the Democratic Republic of Congo.

This extraordinary number of dramatic and historic events with long term consequences demonstrates the significance of this critical three year period in recent history. Here I make just a few astrological correlations with Comet Hale-Bopp.

Tasmania shooting spree

In April 1996 Martin Bryant killed 35 people and injured 23 in the tourist town of Port Arthur, Tasmania, in the worst shooting spree in Australia's history. It resulted in a government ban on private ownership of automatic and semi-automatic firearms. Australia was the first place where the comet was seen with the naked eye and we can see close correlations between the comet's movements and the country's 1788 discovery chart when Captain James Cook arrived in what is now Sydney harbour.

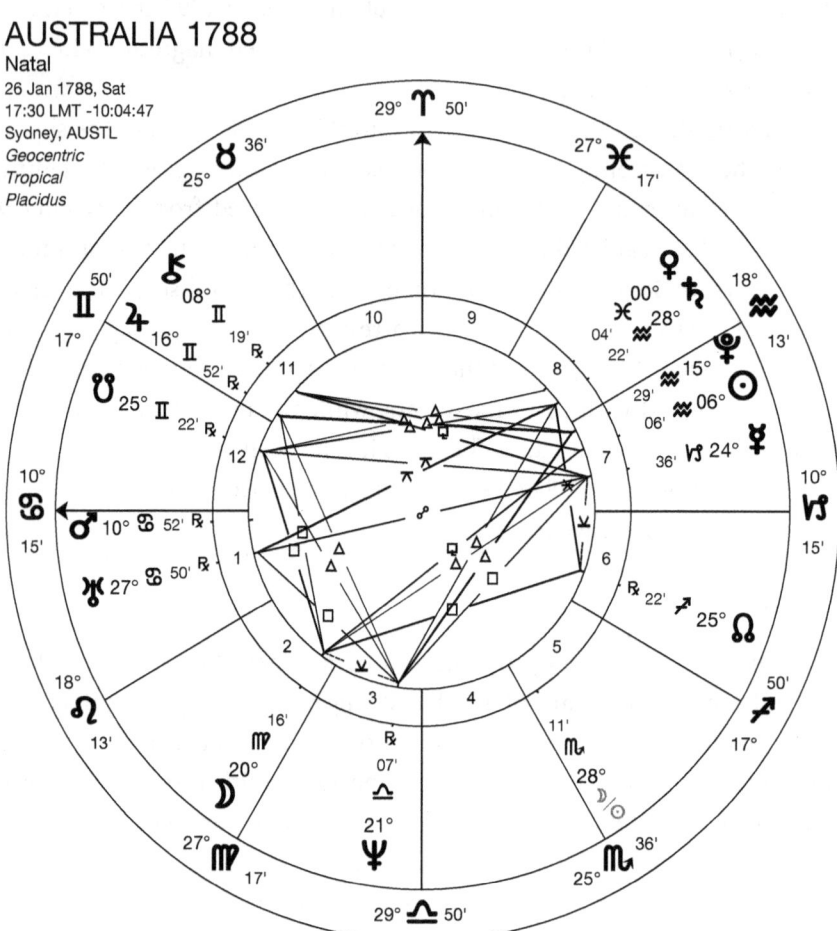

Figure 8.9.1

Comet Hale-Bopp crossed the celestial equator at 9 Cancer, right beside the chart's ASC-Mars conjunction at 10 Cancer. This same degree is where the comet aligned with transiting Mercury on 28 July 1997.

Hale-Bopp was discovered at 29 Cancer, just two degrees from the 1788 chart's Uranus and five degrees from an exact opposition to the chart's Mercury at 24 Capricorn, which itself is just one degree from the point in the sky where the comet was first seen by Lovejoy at 23 Capricorn. Finally, 15 Gemini where the comet crossed the ecliptic is close to Jupiter in the 1788 chart (Figure 8.9.1).

Heaven's Gate

Comet Hale-Bopp played a direct role in the largest ever mass suicide in the US.[130] Cult leader Marshall Applewhite set up the new age cult Heaven's Gate with his partner Bonnie Nettles, persuading his followers that they would ascend into new extraterrestrial bodies by committing suicide when the comet was at its closest to Earth in March 1997.

Just before the suicide of 39 cult members, their website stated: "Hale-Bopp brings closure to Heaven's Gate ...our 22 years of classroom here on planet Earth is finally coming to conclusion – 'graduation' from the Human Evolutionary Level. We are happily prepared to leave 'this world' and go with Ti's crew." Ti was the nickname of his former partner Nettles, who had died of cancer.

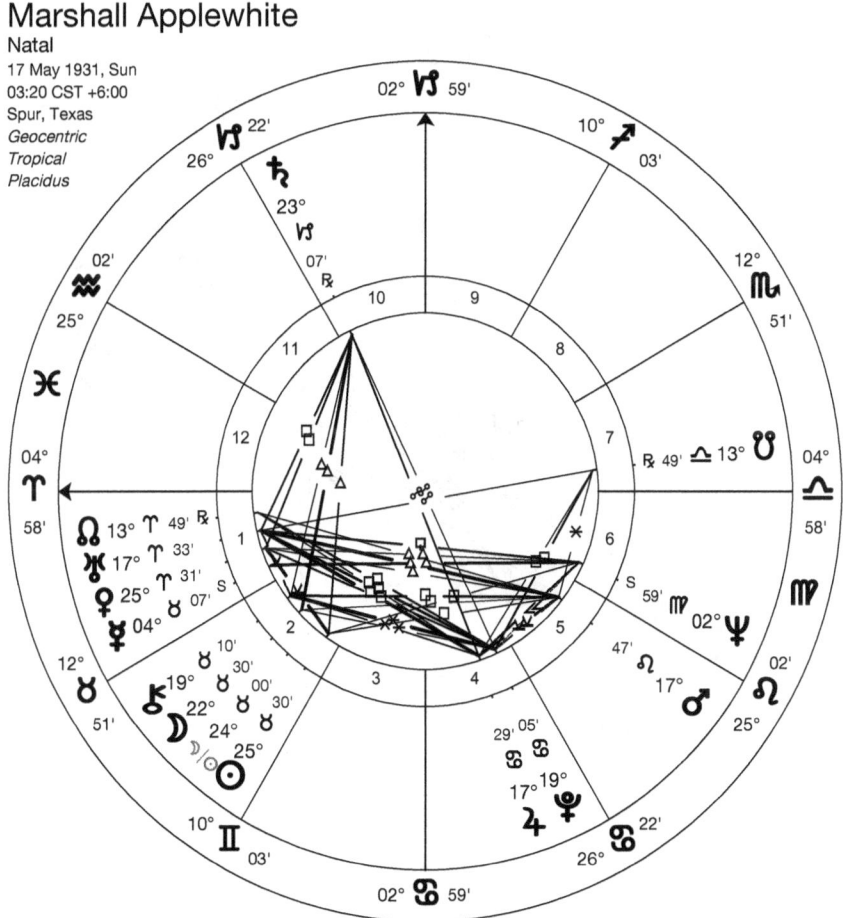

Figure 8.9.2

112   Comets in Astrology

As they prepared the mass suicide, the comet was passing through Aries, home to Applewhite's natal Uranus and Venus. (Figure 8.9.2) His Saturn rules his MC at 23 Capricorn where the comet was first seen. Its transit past the constellation Serpens suggests mass suicide with drugs, as does the alignment with Neptune, and the alignment with Jupiter brings in religion and belief.

Princess Diana

Perhaps the most memorable news event of the period was the highly publicised divorce and death of Princess Diana in a car crash in a tunnel in Paris on 31 August 1997.[131] The comet crossed the celestial equator at 9 Cancer right on her 7th house Sun pointing to a direct connection with her

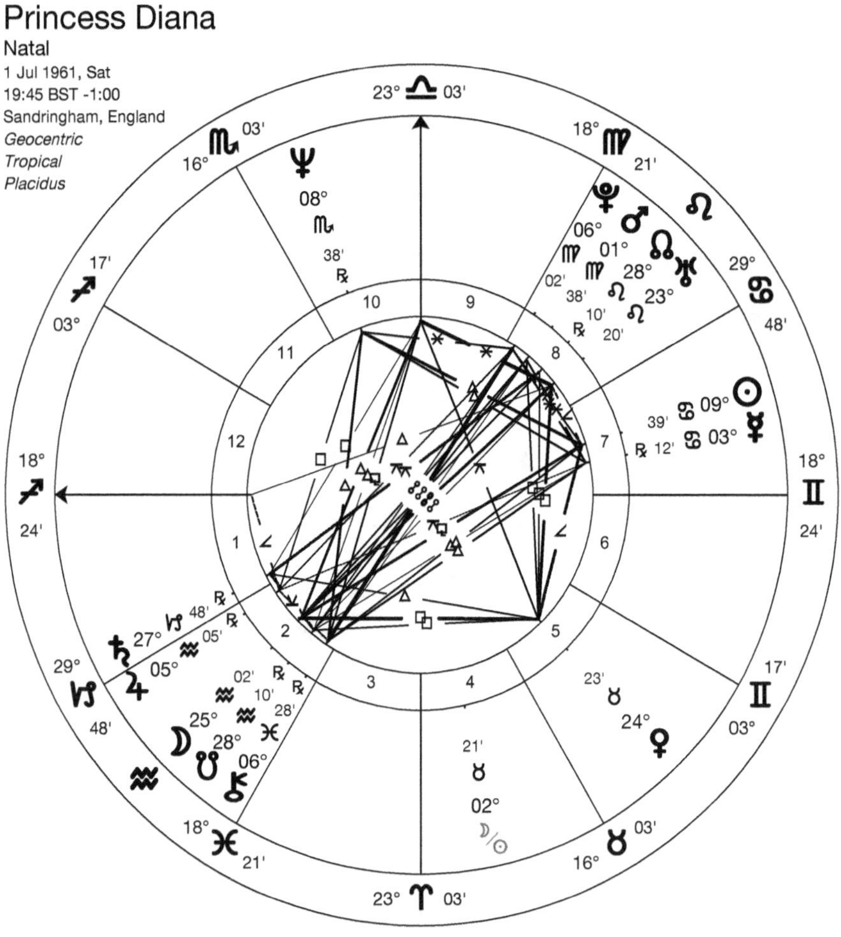

Figure 8.9.3

chart and her relationships. When she died the comet was at 5 Leo, directly opposing her ruling planet Jupiter in Aquarius. (Figure 8.9.3)

Harry Potter

This was also the period when Scottish author JK Rowling published the first of her highly successful Harry Potter books on 26 June 1997.[132] This was when the Comet Hale-Bopp crossed the celestial equator, an important moment in astrology, and may symbolise the enormous impact the magical world she created had on publishing, culture and cinema. And isn't it interesting that on the day of first publication the comet was passing through the constellation Monoceros, the unicorn, whose death in *Harry Potter and the Sorcerer's Stone* came as such a shock.

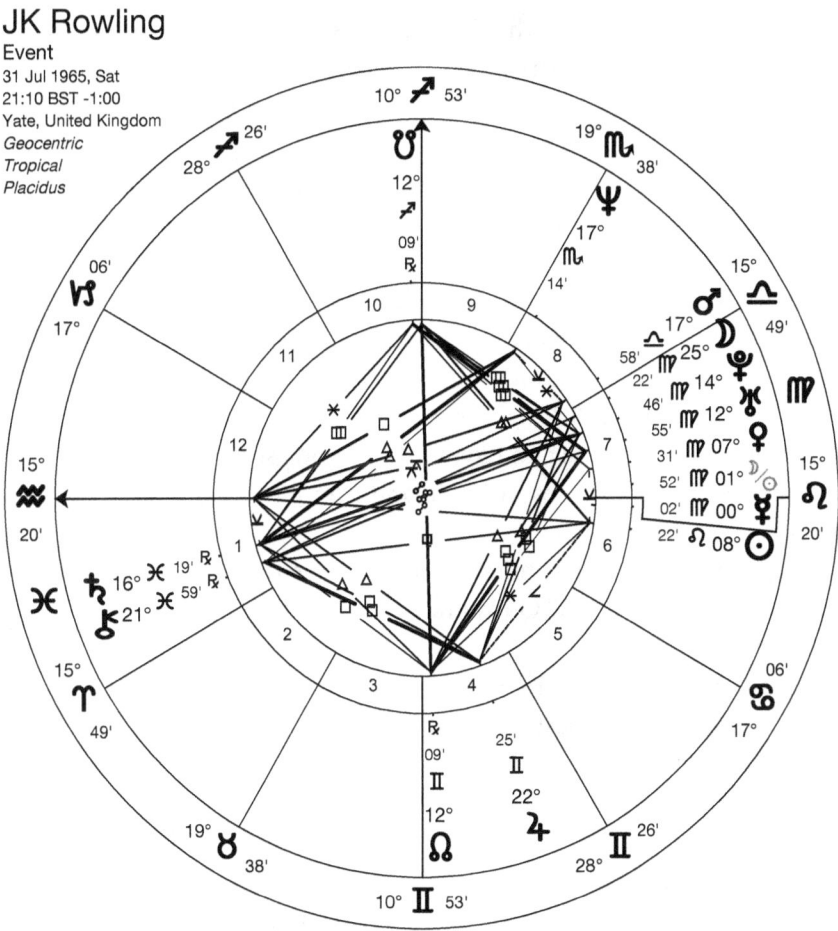

Figure 8.9.4

On 19 November 1997 when Rowling won the Nestlé Smarties Children's Book Prize for her first book *Harry Potter and the Philosopher's Stone*, Comet Hale-Bopp was at 5 Virgo, just two degrees away from her natal Venus at 7 Virgo, having just passed her natal Mercury at 0 Virgo and Sun-Moon midpoint at 1 Virgo.

Comet Hale-Bopp through the Elements

Fire

The unusual number of violent wars, massacres, explosions and fires during this period relate to the comet's passage through all three fire signs, and more specifically its conjunction with the bright star Betelgeuse, symbolic of attacks and the military.

Two wars broke out in the Democratic Republic of Congo, the second being the most deadly since the Second World War with 5.6 million victims by the time it ended in 2003. Wars also broke out in Kosovo, Ethiopia, Guinea-Bissau, Burundi and Liberia and continued in Algeria despite a ceasefire agreement.

The conflict between rebels in Chechnya and the Russian state came to an end after years of fighting. The Troubles in Ireland also ended after 30 years following the reaction to the deadliest bombing in Northern Ireland's history, but the conflict in Sri Lanka between the Tamil Tigers and the government continued. Al Qaeda bombed the US embassies in Tanzania and Kenya killing 224 and injuring 4,500 people as well as staging attacks in Bosnia and Egypt, while Hindu separatists and Brahmin militants staged massacres in India, and over a thousand people died in riots in Indonesia.

The US started firing missiles at Iraq over their refusal to allow in weapons inspectors, NATO expanded eastwards with Poland, Hungary and the Czech Republic all joining and China commenced missile testing and military exercises off Taiwan.

France exploded its last nuclear test bomb, soon followed by the signing of the international Nuclear Test Ban treaty, only for that to be broken by both India and Pakistan.

There were a huge number of major fires and explosions in the US, Brazil, the Philippines, China, Peru, Venezuela, Puerto Rico, Saudi Arabia, Thailand, Finland and Nigeria. Hong Kong had its worst fire with 41 deaths and 81 injuries, two space rockets exploded shortly after lift off and 230 people died when a Boeing 747 exploded off Long Island, New York.

As well as the incident in Tasmania, Scotland saw the Dunblane massacre of schoolchildren and there were school shootings in Mississippi and Arkansas.

Two prominent rappers Tupac and The Notorious B.I.G were shot, as was fashion designer Gianni Versace and Pakistan People's Party leader Murtaza Bhutto.

A massive eruption of the Soufrière Hills volcano forced Montserrat capital Plymouth to be evacuated.

Earth

The earth element relates to money, matter and the body and during this period we saw the founding of the European Central Bank and the first Euro coins being minted in France.[133] The Russian financial crisis of 1998 was spawned when the government devalued the ruble, defaulted and declared a moratorium on repaying foreign debt after being loaned $10.2 billion by the International Monetary Fund.[134] The Asian financial crisis was also triggered by the Bank of Thailand. Meanwhile Disney purchased ABC's parent company for $19 billion and the largest merger in US history took place between WorldCom and MCI Communications worth $37 billion.

There were major earthquakes with massive casualties in China, Indonesia, Iran, and in Afghanistan where two consecutive earthquakes killed more than 1,500 people and injured over 4,000. An earthquake off Papua New Guinea triggered a tsunami that killed 2,100 people and injured thousands more.

A huge biotech breakthrough occurred with the successful cloning of Dolly the sheep, 'born' in Scotland in July 1996 and sparking major ethical debates, leading to a European ban on funding human cloning research. This was also when the first three parent baby was conceived in New Jersey through mitochondrial donation.

Air

The period saw a huge number of airplane accidents in Congo, the Dominican Republic, Croatia, Peru, Brazil, Nigeria, India, Ethiopia, Israel, Turkey, Guam, Indonesia, Uruguay, Italy, Taiwan, Canada and Thailand, with a combined death toll of 2,624. A train crash in China killed 126 people and 101 died when a high speed train derailed in Germany. Long standing Dutch aircraft manufacturer Fokker went bankrupt and Kuala Lumpur international airport was opened.

In space an unmanned Russian spacecraft collided with the *Mir* space station, the first piece of the International Space Station was launched by a Russian rocket from Kazakhstan with a second piece launched soon afterwards from the US Space Shuttle. Japan became the third nation to explore outer space launching a space probe to Mars. The *Lunar Prospector* discovered frozen water on the Moon, the *Galileo* probe found liquid ocean under the ice crust on Jupiter's moon Europa, evidence was published showing the universe is expanding at an accelerating rate and the first space burial took the remains of 24 people into orbit on board a *Pegasus* rocket.

This was a period of extreme weather events with the second deadliest hurricane in history killing 11,374 people in Central America, while Severe Tropical Storm Linda killed 3,275 people in Vietnam and Thailand, at least 1,000 died in Andra Pradesh, India, in a category 4 cyclone, and 600 were killed in severe storms in Bangladesh. Hurricanes cost billions in damage and loss of life in North Carolina, hundreds died in a US ice storm and Texas had one of its deadliest tornados.

Meanwhile world leaders gathered in Japan on 11 December 1997 as Comet Hale-Bopp disappeared from sight to sign the Kyoto Protocol to tackle climate change by reducing greenhouse gas emissions.

In 1996 the highly pathogenic avian influenza strain H5N1 was detected in southern China, spreading further afield the following year and fear when it was discovered this virus could infect humans and cause the death of children.[135] We can also relate this to the comet passing through the constellation of Aquila and Cygnus in early 1997. Hong Kong killed 1.25 million chickens to stop the spread of the virus.

Another air connection came with IBM's Deep Blue supercomputer defeating world champion chess master Gary Kasparov.[136] I have noticed major breakthroughs in brain and computer research when comets appear. Netflix was founded as a DVD-by-mail rental service, first person shooter video game *Half-Life* was launched, and remains one of the most influential ever, and the *New York Times* published its first front page colour photograph.

And probably most significant of all, Google was founded by Larry Page and Sergey Brin, and the commercial internet started to spread into homes around the world, fundamentally changing the nature of human interaction.[137]

Water

When the comet passed through Cancer in July 1997 central Europe experienced widespread flood damage with 114 lives lost. Also in this three year period heavy rains killed 80 campers in Spain, we had Africa's worst boat disaster with nearly a thousand people dying when their boat capsized on Lake Victoria in Tanzania, an Indonesian ferry sank off Sumatra killing more than 100 people, 81 drowned east of Uganda when their boat capsized, a boat carrying market traders capsized in Sierra Leone killing 86, and 283 migrants drowned off Sicily.

Canada experienced its worst ever flood, floods in North Dakota and Minnesota cost billions, but worst of all were the 1998 Yangtze River floods in China that affected an estimated 180 million people. More than 4,000 people are thought to have died, 15 million were made homeless and the economic loss was $24 billion.

Other events of significance include:

- scientists published DNA research demonstrating humanity was descended from an African "Eve" who lived up to 200,000 years ago;
- China, Kazakhstan, Kyrgyzstan, Tajikistan and Russia formed the Shanghai Five Group that evolved into the Shanghai Co-operation Organisation, creating a powerful political, economic and defence alliance;
- The African National Congress assumed full political control in South Africa for the first time;
- Benjamin Netanyahu narrowly won his first election victory in Israel for Likud;
- The Czech Republic held its first election;
- Ukraine's constitution was signed into law;
- Yoweri Museveni won a landslide victory in Uganda's first presidential election;
- The Taliban took control in Afghanistan;

- Tony Blair won the UK election for Labour for the first time in 18 years, started the peace process in Northern Ireland and the move to devolution for Scotland, Wales and Northern Ireland;
- China took over sovereignty of Hong Kong;
- Madeleine Albright became the first female US Secretary of State;
- Buddhist and Christian monks, including the Dalai Lama, held their first intermonastic dialogue;
- In the US the Monica Lewinsky scandal led to Bill Clinton becoming only the second president ever to be impeached;
- After years of pressure, Iraqi president Saddam Hussein allowed weapons inspectors in;
- Suharto resigned as Indonesia's president after 31 years in power;
- The International Criminal Court was established and in the first ever conviction for inciting genocide, Rwandan mayor Jean-Paul Akayesu was sentenced to life imprisonment for his role in the 1994 genocide of the Tutsis;
- North Korea enshrined Kim Il Sung as eternal president, creating the country's military dictatorship;
- Iran's president Mohammad Khatami rescinded the fatwa against novelist Salman Rushdie, only for it to reapplied seven years later;
- Europol, the first Europe-wide police force, was created; and
- Hugo Chavez was elected president of Venezuela.

Comet McNaught (C/2006 P1)

Type of Comet: non-periodic

Astronomical data

	Date	Position
Discovery	07 Aug 2006	10 Sagittarius
First seen	01 Jan 2007	10 Capricorn
Last perihelion	12 Jan 2007	26 Capricorn
Last seen	24 February	
Crossed ecliptic	14 Jan 2007 01:19 UT	28 Capricorn

Tropical Zodiac

Date	Sign
01 Jan 2007 - 15 Jan 2007	Capricorn
15 Jan 2007 - 28 Feb 2007	Aquarius

Constellations

Note: not to be confused with the 12 signs of the tropical zodiac.

Date	Constellation
01 Jan 2007 - 04 Jan 2007	Scutum
04 Jan 2007 - 11 Jan 2007	Aquila
11 Jan 2007 - 18 Jan 2007	Sagittarius
18 Jan 2007 - 24 Jan 2007	Indus
24 Jan 2007 - 28 Feb 2007	Tucana

Planetary alignments

Date	Planet	Zodiac location
01 Jan 2007	Sun	10 Capricorn
14 Jan 2007	Mercury	29 Capricorn
23 Jan 2007	Sun	01 Aquarius

Comet McNaught, the Great Comet of 2007, was discovered by Scottish-Australian astronomer Robert H McNaught on 7 August 2006[138] and became visible to the naked eye in the Southern Hemisphere for a few weeks the following January and February, reaching a peak magnitude of -5.5.

Comet McNaught stood out because of its brightness and its extraordinary tail, shaped like a fan or a broom, the symbolic meaning of which implies waves, the wind, the air or cleaning things up, even wiping things out. Its colour was orange and gold, suggesting the nature of the Sun.

On 1 January 2007 it had reached 6th magnitude making it bright enough to be seen without technical assistance[139]. At the time it was passing through the constellation of Scutum, the shield, symbol of protection and defence.

Between 5 and 11 January it passed through Aquila, the eagle, suggesting themes of attack and military action.

Between 12 and 14 January, and again between 17 and 18 January, the comet travelled through the constellation of Sagittarius, signifying philosophy, religion, publishing, higher education, long distance travel and flight.

From 19 to 23 January it moved through the constellation of Microscopium, highlighting themes of science and the study of small things.

For one month from 24 January until it disappeared from view on 24 February it stayed in the constellation of Indus, which may relate to the indigenous culture of the Americas or any other part of the world for that matter.

Finally it moved into the constellation of Tucana before disappearing for good.

The zodiac signs it crossed were Capricorn, the sign of authority and structure, and Aquarius, the sign of technology, community and rebellion.

It aligned with the Sun at 10 Capricorn the day it was first seen without aid, suggesting an influence on celebrities and leaders. It was then aligned with Mercury at 29 Capricorn on 15 January, bringing up issues of communication and transport, young people and trade.

Events

By far the most significant event, or series of events, that occurred during the passage of Comet McNaught through the sky relates directly to the sign in which it was first observed.

As we know Capricorn relates to structure, government and large institutions, especially financial ones that wield great power. The comet was first seen conjunct the Sun when Pluto was just a couple of degrees from its first entry into the sign the following January.

And this was the period now described as the Great Financial Crisis, that impacted the entire world, shaking the very foundations of the global financial system, bringing the biggest banks to the very brink of bankruptcy and pushing some, like Lehman Brothers, over the edge into collapse.

The first indications that the sub prime mortgage scam would bring the financial world to its knees were felt within six months of the aptly named Comet McNaught's appearance. A year later governments were bailing the banks out, impoverishing nations and bringing austerity to societies all over the world. The effects are still being felt almost 20 years later.

Symbolically this was also the period when Bernard Madoff was convicted of the biggest Ponzi scheme in history worth $65 billion, and it also saw the publication of the white paper announcing the launch of Bitcoin, the world's first and most successful decentralised cryptocurrency.

From Capricorn the comet moved into Aquarius where it spent more than a month before disappearing from view once and for all. If we relate technology to the sign of the Water Bearer, we can see some significant developments:

- The launch of the first smartphone, Apple's iPhone, soon followed by Apple's AppStore and the Android smartphone;
- The launch of the microblogging site Twitter, now called X;
- Google purchased the video platform YouTube for $1.65 billion;
- *Grand Theft Auto IV* became the highest grossing video game and entertainment product ever released;
- The founding and first revelations from whistleblower site Wikileaks, which came to prominence in 2010 with its leaked video "Collateral Murder" taken by a US chopper crew killing civilians and reporters in Iraq, an event that occurred in 2007 around the time of Comet McNaught.

On the scientific front we saw the Human Genome Project publish the final sequence of human chromosomes and the launch of the Large Hadron Collider at CERN, in Switzerland, both of which also reflect the constellation Microscopium - things don't get much smaller than genes and electrons.

There were major earthquakes, with 87,000 people killed in China in May 2008, and major weather events as well with 138,000 people killed the same month when Cyclone Nargis hit Myanmar.

We saw women break through the glass ceiling to become president in Chile and Argentine, Nancy Pelosi became the first female Speaker of the House in the US, and the first female Governor General of Australia was appointed. Barack Obama became the first black US president and Lewis Hamilton became the first black Formula One motor racing champion.

The alignment with Aquila correlates with an intensification of various conflicts around the world, especially in the Middle East with the election of Hamas in Gaza leading to Israel invading the territory, having already bombed Syria and invaded Lebanon to start a war with Hezbollah.

A civil war erupted in Iraq with the bombing of a mosque and after former leader Saddam Hussein was executed there was a series of car bombs that killed hundreds of people. There were terrorist attacks in India and Pakistan, the drug wars in Mexico erupted as did the war between Russia and Georgia, becoming Europe's first 21st century land war.

The alignment with the constellation Indus correlates with events affecting indigenous communities with the election of the first indigenous leader in the Americas, Evo Morales, in Bolivia, the signing of the UN's Declaration on the Rights of Indigenous People, Canada apologising for the Indian residential school system and Australian prime minister Kevin Rudd delivering a formal apology to the Stolen Generations.

Also in 2008 Mauritania became the last country in the world to criminalise slavery, a theme that has often appeared alongside the appearance of comets in our skies.

Comet Pan-STARRS (C/2011 L4)

Type of Comet: non-periodic

Astronomical data

	Date	Position
Discovery	11 Jun 2011	03 Sagittarius
First seen	07 Feb 2013	23 Capricorn
Perihelion	10 Mar 2013	03 Aries
Crossed celestial equator	12 Mar 2013 02:30 UT	06 Aries
Crossed ecliptic	13 Mar 2013 10:05 UT	08 Aries

Tropical Zodiac

Date	Sign
07 Feb 2013 - 11 Feb 2013	Capricorn
11 Feb 2013 - 24 Feb 2013	Aquarius
24 Feb - 08 Mar 2013	Pisces
08 Mar - 12 Apr 2013	Aries

Constellations

Note: not to be confused with the 12 signs of the tropical zodiac.

Date	Constellation
07 Feb 2013 - 10 Feb 2013	Telescopium
10 Feb 2013 - 16 Feb 2013	Microscopium
16 Feb 2013 - 23 Feb 2013	Grus
23 Feb 2013 - 26 Feb 2013	Piscis Austrinus
26 Feb 2013 - 02 Mar 2013	Sculptor
03 Mar 2013 - 05 Mar 2013	Aquarius
05 Mar 2013 - 09 Mar 2013	Cetus
09 Mar 2013 - 22 Mar 2013	Pisces
22 Mar 2013 - 09 Apr 2013	Andromeda
09 Apr 2013 - 30 Apr 2013	Cassiopeia

Fixed Stars

Date	Star
17 Mar 2013	Algenib
22 Apr 2013	Caph

Planetary alignments

Date	Planet	Zodiac location
25 Feb 2013	Venus	29 Aquarius
26 Feb 2013	Neptune	03 Pisces
28 Feb 2013	Chiron	09 Pisces
28 Feb 2013	Sun	10 Pisces
02 Mar 2013	Mercury R	15 Pisces
07 Mar 2013	Mars	26 Pisces
12 Mar 2013	Uranus	07 Aries

Comet Pan-STARRS holds special significance for myself as an astrologer, as this comet inspired my search for the astrological meaning behind the appearance of these messengers in the sky. When a reporter asked me to explain what its appearance symbolised, I was ashamed that all I could say was "disaster". His inquiry triggered my journey researching all I could about comets and publishing the article 'Comets in Astrology' in *The Astrological Journal* in 2014.

Comet Pan-STARRS was first observed on 7 February 2013 conjunct my MC ruler Jupiter before aligning with transiting Mars at 26 Pisces, the degree of my MC!

The comet became visible for almost three months at the beginning of 2013 as it travelled through the signs of Capricorn, Aquarius, Pisces and Aries from 7 February until late April, thus reflecting all four of the elements earth, air, water and fire. This was a year of multiple comets, which was reflected on Earth by the number of significant and extreme events with long term implications. It also coincided with the Uranus-Pluto square and no doubt enhanced the intensity of the events this powerful configuration would signify.

Events

Edward Snowden

2013 was the year that Edward Snowden hit the headlines with his revelations about the covert surveillance programme being operated by his former employer, the US National Security Agency, in conjunction with western intelligence agencies and telecommunications companies. After his concerns were ignored by his employers, he fled to Hong Kong and leaked thousands of classified documents to selected journalists, before seeking asylum in Russia. He was charged in the US with espionage, but his

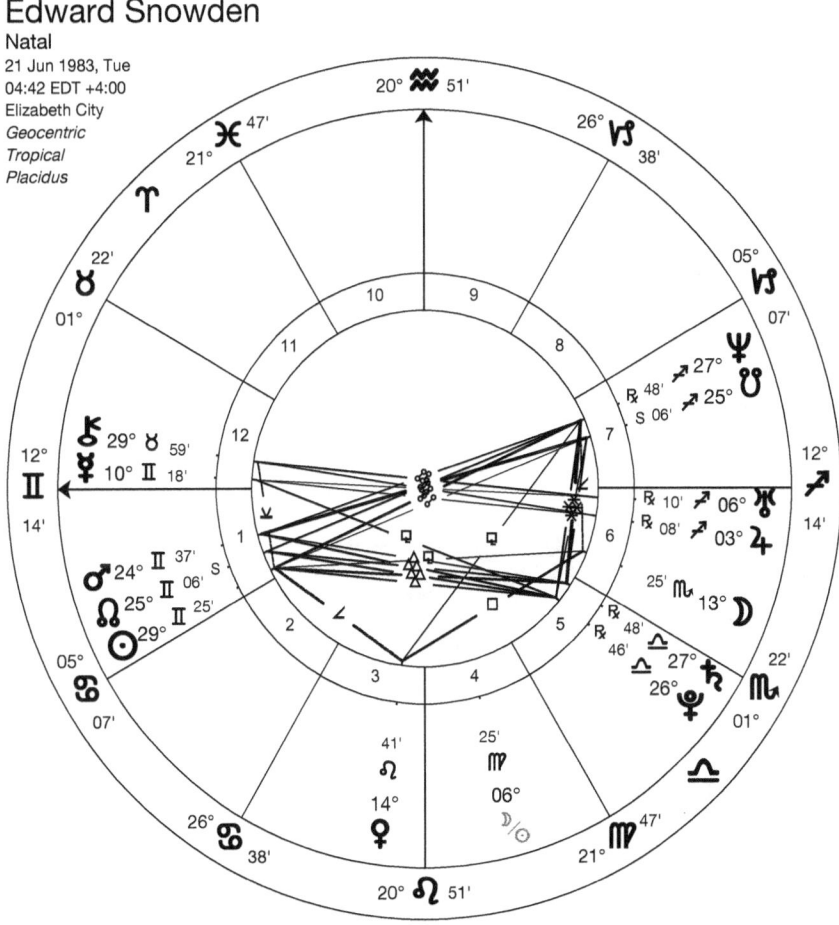

Figure 8.11.1

126 Comets in Astrology

actions exposed government secrecy and sparked international discussion of an individual's right to privacy.

Snowden's Jupiter-Uranus conjunction opposing his Mercury-Ascendant is the hallmark of the archetypal whistleblower, but we can note that Comet Pan-STARRS was first discovered on 11 June 2011 at 3 Sagittarius, the degree of his natal Jupiter, which may well have been when his concerns first arose. (Figure 8.11.1)

Pope Francis

Another big shock that year, this time in the world of religion, was the sudden resignation of Pope Benedict XVI. He was the first pontiff to resign since the forced departure of Gregory XII back in 1415, and the first to

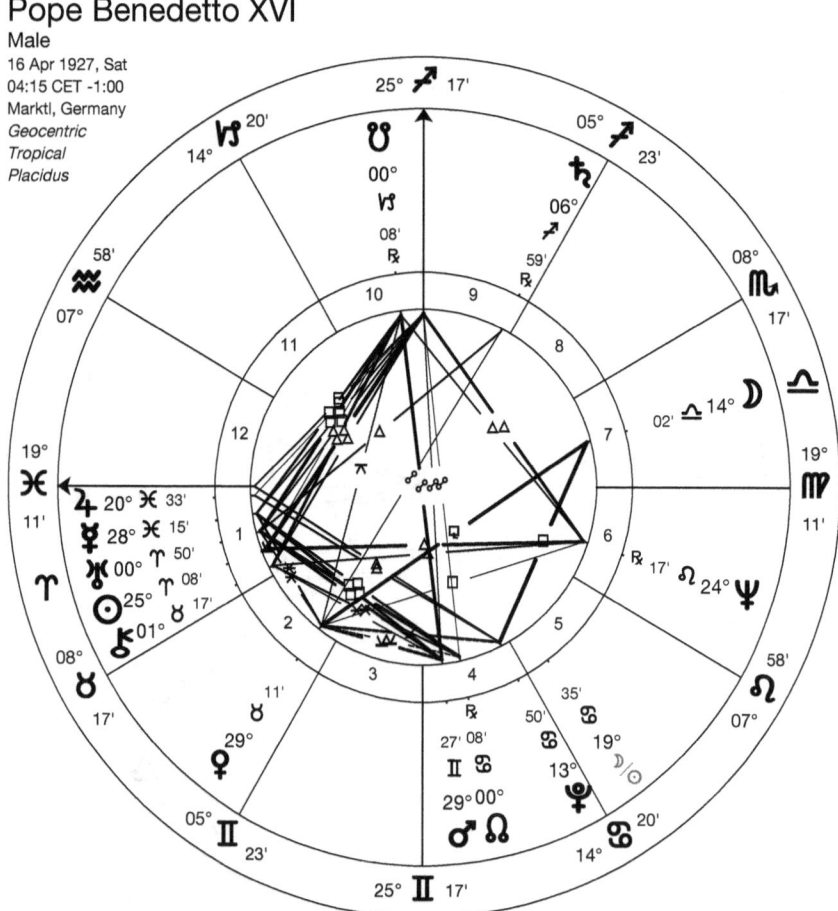

Figure 8.11.2

resign voluntarily since Celestine V in 1294. All other popes have held the position until they died. (Figure 8.11.2)

Once again, the day of the comet's discovery at 3 Sagittarius impacted Benedict's birth chart, being the degree of his natal Saturn and square Chiron-Neptune in the comet's discovery chart. When the comet was first seen on 7 February at 23 Capricorn it was opposite his natal Mars. On 25 February 2013 the comet passed Venus at 29 Aquarius and entered Pisces to align with transiting Neptune and then Chiron on 28 February, the day of his resignation. This would reflect the confusion and rumours surrounding his announcement, which he insisted was due to poor health.

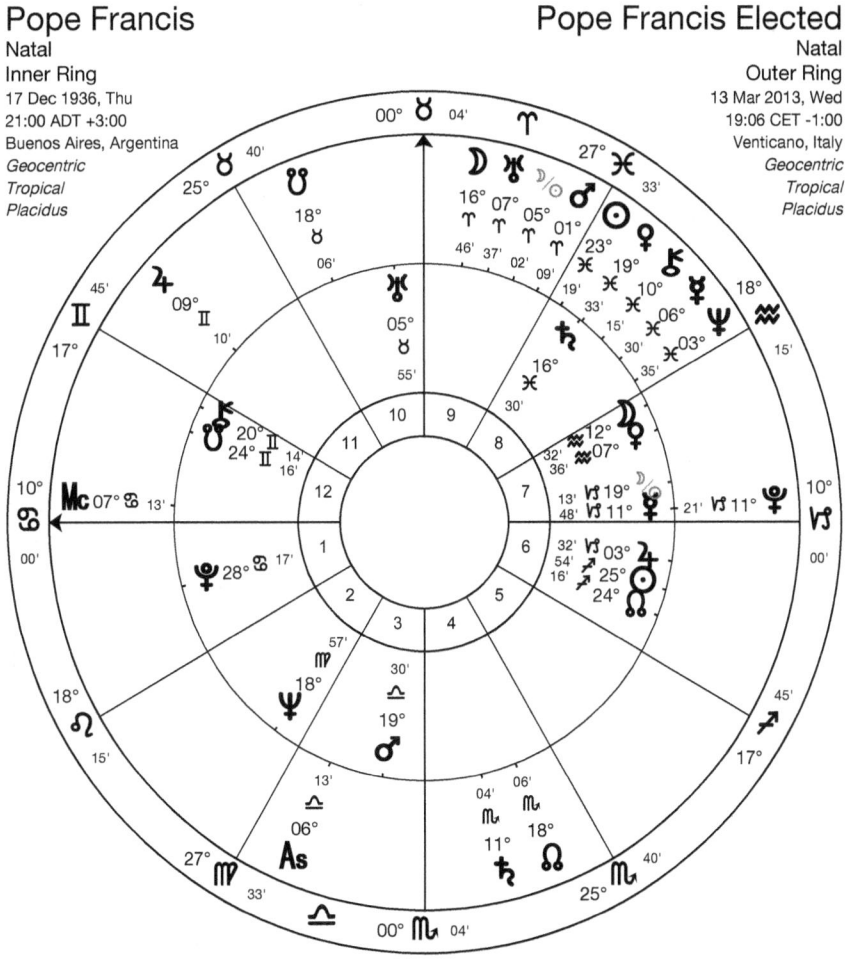

Figure 8.11.3

The day of Benedict's announcement on 11 February was one day after the Aquarius new Moon when Uranus was on the Vatican's Ascendant at 6 Aries, a reflection of the shock the news provoked.

His successor Pope Francis was elected on 13 March, the day the comet reached its brightest and crossed the ecliptic to align with Uranus at 8 Aries. Two days earlier the comet had aligned with the new pope's natal Saturn at 16 Pisces as it passed the new Moon in the sky having just aligned with Mercury on 7 March. (Figure 8.11.3)

Pope Francis represented a break with tradition of the kind that comets often bring. He was both the first Jesuit to become pope and the first from the southern hemisphere. He has allowed women to hold more prominent positions in the Catholic church, and though he has not endorsed gay marriage or transgenderism, he told a televised press conference on 28 July 2013: "If someone is gay and is searching for the Lord and has good will, then who am I to judge him?" His statement was widely reported and helped build his image as a progressive reformer.

Abdications

Comet Pan-STARRS' appearance also coincided with sudden, unexpected changes in leadership. On 5 March the hugely controversial and influential Venezuelan president Hugo Chavez, who birthed the Bolivar revolution that influenced politics throughout South America, died shortly after he had won his fourth term aged just 58.

Just as the comet was leaving the constellation Cassiopeia, the queen of heaven, in April 2013, Queen Beatrix of the Netherlands abdicated in favour of her son Willem-Alexander.

The following year King Juan Carlos I abdicated, leaving the Spanish throne to his son Felipe VI, after apologising for going on a controversial elephant hunting safari in Botswana. He was also forced to resign from his position as honorary president of the Spanish branch of the Worldwide Fund for Nature.

Comet Pan-STARRS aligned with the king's MC at 23 Capricorn when it first appeared, passed by his Moon at 25 Aquarius before aligning with transiting Venus at 29 Aquarius close to his natal Moon. It then aligned with the Sun at 10 Pisces, conjunct his natal Mars on 7 March, and with transiting Mars at 26 Pisces close to natal Saturn, his MC ruler. (Figure 8.11.4)

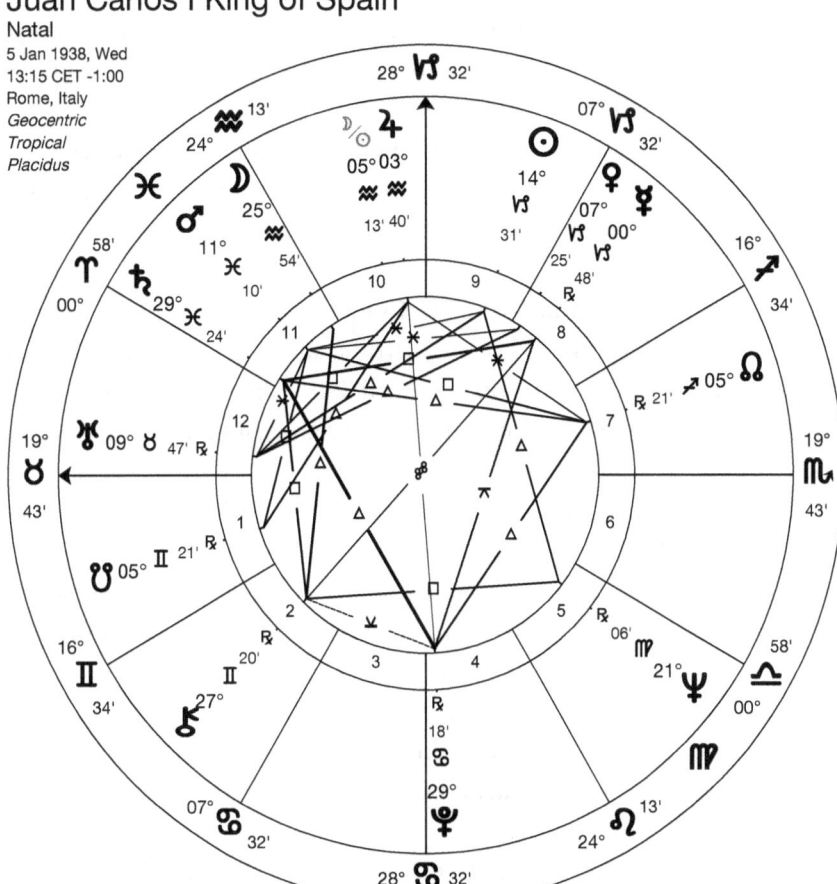

Figure 8.11.4

July 2013 saw a military coup under the leadership of General Abdel Fattah el-Sisi to depose Egypt's first democratically elected president Mohamed Morsi. Large scale protests the following month against the military saw between 900 and almost 2,700 protesters killed in the largest massacre of demonstrators in modern history.[140] The previous year military coups had taken place in Mali and Guinea-Bissau, while in 2014 a coup took place in Thailand.

The seeds were sown that led to the war in Ukraine between Russia and the West in this period when the Euromaidan uprising overthrew pro-Russian president Viktor Yanukovych after he decided to shift the country's allegiance towards Russia rather than Europe. The move led to

Russia's annexation of Crimea and residents in Ukraine's eastern territories of Donetsk and Luhansk to vote for independence, which led to civil war. When Comet Pan-STARRS was discovered at 3 Sagittarius it was right on Mars in Ukraine's chart.

This period also saw civil wars break out in the Central African Republic, Sudan and Yemen, major protests in Hong Kong, an escalation of Israel's war on Gaza and ISIS declaring a caliphate that led to massacres in both Iraq and Syria.

There were terror attacks in the Congo, Pakistan, England, Kenya, China, Nigeria, Taiwan, Mexico, Australia and the US, including the Boston Marathon bombing on 15 April 2013 shortly after the comet disappeared having passed through the fire sign Aries and aligned with a Mars-Uranus conjunction.

The period also saw new epidemics, first with the Middle East respiratory syndrome (MERS), then a polio health emergency in Pakistan and worst of all the Ebola virus epidemic that infected 28,646 people and led to the death of 11,323 sufferers in West Africa, all of which may link to the comet's alignment with Neptune.

Pan-STARRS' passage through the earth sign Capricorn is reflected in the Greek debt and Cyprus banking crises, major earthquakes in the Philippines, Indonesia, Alaska, and China where 11,000 were injured and 193 died in Sichuan province.

A week after the comet was first observed and as it travelled through the air sign Aquarius, a meteor exploded above the Russian city of Chelyabinsk, injuring around 1,500 people and damaging over 4,300 buildings. It was the most powerful meteor to strike the Earth's atmosphere for more than a century and happened while the comet aligned with Russia's Saturn at 6 Aquarius.

The air element was also represented by some of the worst weather events ever recorded, including the 900 mile wide Hurricane Sandy that struck the Caribbean and the US east coast in late 2012, killing 233 people and causing damage worth $68.7 billion. The following May another record was broken by the 2.6 mile wide El Reno tornado that reached wind speeds over 300mph and killed eight people in its path. Then in November Vietnam and the Philippines experienced Super Typhoon Yolanda, one of the most powerful tropical cyclones ever recorded, that led to 6,241 deaths.

There was also no shortage of aircraft disasters, with Malaysia Airlines worst affected. In March 2014 Flight 370 disappeared over the Gulf of Thailand with 239 people on board, while just four months later Flight 17 was shot down by a missile over eastern Ukraine with the loss of all 298 passengers and crew. We may note that as well as passing through air sign Aquarius, the comet aligned with fixed star Algenib in the constellation Pegasus, the symbol of flight.

Looking at the midnight chart of 1 May 1947, the day the airline was founded, the Ascendant is at 23 Capricorn, the degree where the comet was first observed by the naked eye. It had been discovered at 3 Sagittarius, on the airline's South Node and reached perihelion, crossed the celestial

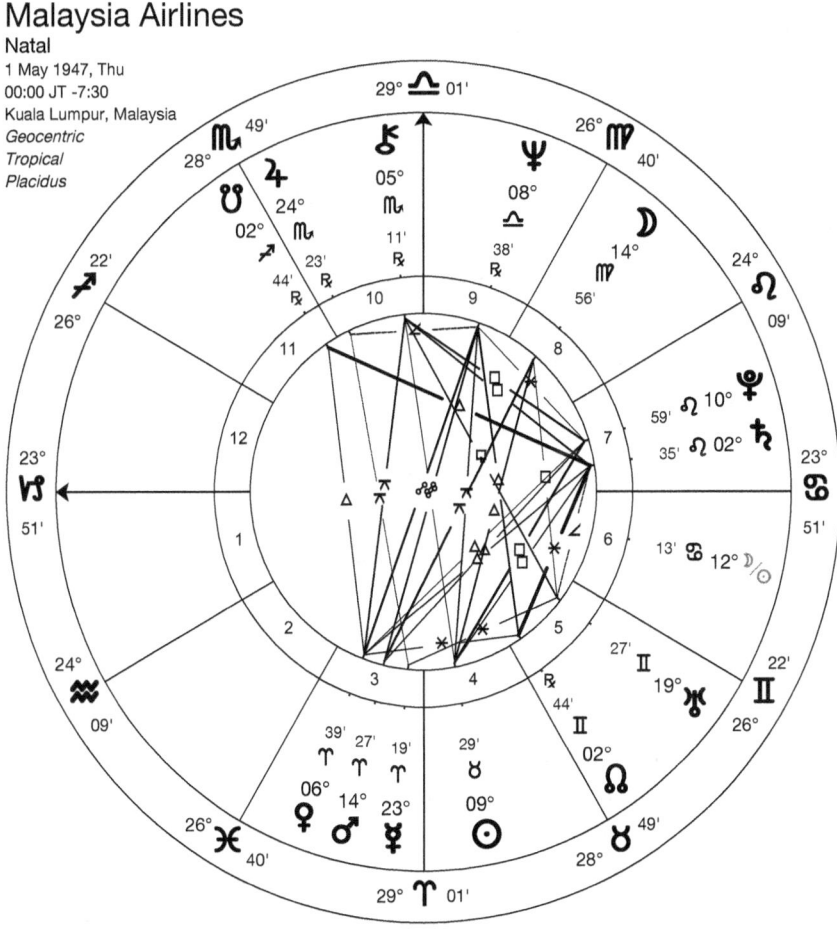

Figure 8.11.5

equator and the ecliptic in early Aries close to the company's MC ruler Venus. (Figure 8.11.5)

Incidents related to the element of water also abounded during this turbulent period under the influence of Comet Pan-STARRS. In September 2013 the Colorado River Basin had a year's worth of rain in just four days flooding a 200 mile stretch of the river, leaving eight dead, two missing presumed dead and hundreds unaccounted for. 11,000 people had to be evacuated, 19,000 homes were damaged with more than 1,500 completely destroyed and the total cost of the damage estimated at more than $1 billion.

Four months later in January 2014, 38,000 litres of the chemical methylcycloexanemethanol (MCHM) used for washing coal, leaked into the Elk River, in Charleston, West Virginia, leaving 300,000 people without potable water. Fourteen were taken to hospital with symptoms, but a further 170 were affected by nausea.

Then in April, South Korea experienced its greatest shock since the Korean War in the 1950s when the ferry MV Sewol capsized and sank off the peninsula's south west coast when its cargo unexpectedly shifted. While the captain and crew escaped, 304 passengers including 250 high school students were told to stay put and ended up drowning while rescue operations were stalled. The news sent shockwaves throughout East Asia and fueled campaigns against government corruption and incompetence, which were still ongoing a decade later with a 10th anniversary march on the government's headquarters in Seoul that lasted 21 days.

In science the world saw several breakthroughs in artificial organ technology, with the first 3D printed ear, FDA approval of the Argus II bionic eye and the first cloning of human embryonic stem cells.

Scientists at CERN in Switzerland confirmed their Large Hadron Collider, which opened around the time of Comet McNaught, had identified the long sought after Higgs Boson particle. Meanwhile in northern Chile the Atacama Large Millimeter Array, the world's most powerful radio telescope, became fully operational.

It is worth noting that the comet was first observed while passing through the constellation Telescopium, whose symbol is the telescope, before moving through Microscopium, whose symbol is the microscope.

Comet Wirtanen (46P)

Type of comet: periodic

Orbital period: 5.44 years

Astronomical data

	Date	Location
Discovery	17 Jan 1948	11 Gemini
First seen	Late Nov 2018	17-22 Aries
Last perihelion	12 Dec 2018	20 Taurus
Last seen	Late Jan 2019	0-4 Leo
Crossed celestial equator	10 Dec 2018 04:37 UT	16 Taurus
Crossed ecliptic	16 Dec 2018 01:06 UT	01 Gemini

Tropical Zodiac

Date	Sign
20 Nov - 01 Dec 2018	Aries
01 Dec - 15 Dec 2018	Taurus
15 Dec - 31 Dec 2018	Gemini

Constellations

Note: not to be confused with the 12 signs of the tropical zodiac.

Date	Constellation
20 Nov - 27 Nov 2018	Fornax
27 Nov - 04 Dec 2018	Cetus
04 Dec - 09 Dec 2018	Eridanus
09 Dec - 11 Dec 2018	Cetus
11 Dec - 18 Dec 2018	Taurus
18 Dec - 21 Dec 2018	Perseus
21 Dec - 27 Dec 2018	Auriga
27 Dec - 31 Dec 2018	Lynx

Fixed Stars

Date	Star
13 Dec 2018	Aldebaran

Planetary alignments

Date	Planet	Location
01 Dec 2018	Uranus	29 Aries

Comet 46P/Wirtanen was one of our brightest comets when it appeared at the tail end of 2018, reaching magnitude 3 on 16 December after becoming visible from the end of November. It traversed the zodiac signs Aries, Taurus and Gemini between November and December 2018, aligning with just one planet, Uranus.

Ancient astrologers related bright comets to shocking events, such as natural disasters and the death of prominent figures. My research suggests bright comets also correlate with groundbreaking inventions, important technological breakthroughs, developments in communications and transport, and 'release' in all the myriad ways that word could express itself.

At the time it crossed the heavens I wrote that its alignment with Uranus would "inevitably bring forth some shocking news related to the energy industry, advanced technology and financial markets", as well as protests, riots and earthquakes. I suggested its passage through Taurus would bring up issues around personal finance and nature, and through Gemini would affect students, transport and communications.

Events

Protests

Looking first at the Uranus alignment, these two years saw scenes of enormous protests throughout the world that grew in intensity.

The seeds of two of the most prominent protest movements that went on to take firm root in society - the Yellow Vests in France and Extinction Rebellion - were sown while Comet 46P appeared in the sky.

The Yellow Vest movement hit the streets of French towns and cities on 17 November 2018 triggered by anger about rising fuel costs, with protesters wearing the yellow high-visibility vests French law made every driver have in their car in case of an emergency.

The movement grew in size rapidly after 300,000 took part on that first day, and within weeks over a million people were out on the streets every weekend. It also became increasingly violent as the police response escalated, with hundreds of casualties, including deaths. Within a few months around 2,000 protesters had been injured along with more than 1,000 police officers, and 11 people had lost their lives.

The protests continued every weekend for over a year and only stopped with the Covid 19 lockdown in 2020. They continued once the lockdown was lifted, though not on quite the same scale. However the movement was picked up and mirrored internationally with yellow vest protests taking place in no less than 24 other countries.

When the comet aligned with Uranus on 1 December 2018 at 29 Aries, it was closely conjunct natal Saturn in the chart of France's 1st Republic,

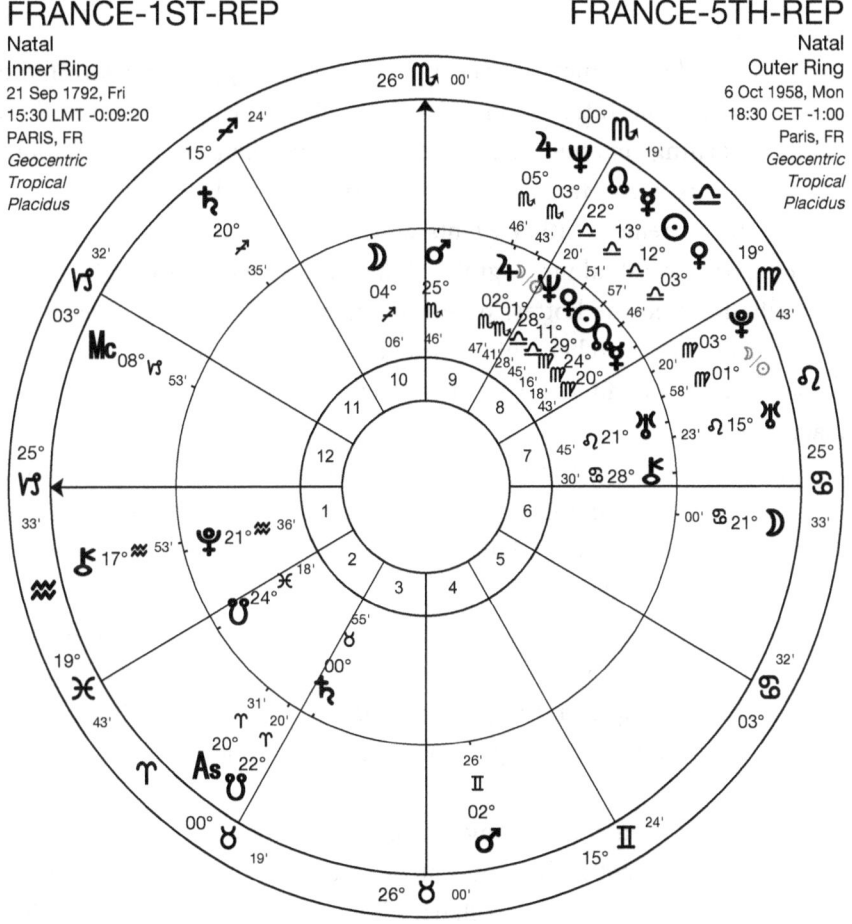

Figure 8.12.1

and it crossed the ecliptic at 1 Gemini right on top of natal Mars in the chart of the country's 5th Republic. (Figure 8.12.1).

Extinction Rebellion was founded in October 2018 and staged its first major protest outside London's Houses of Parliament on the last day of that month demanding action to tackle "climate breakdown". The protest grew in strength and size turning into a global movement with non-violent direct actions taking place throughout the world as concern spread about climate change.

Swedish school student Greta Thunberg had begun her school strike for the climate in August 2018 aged just 15 and within six months had addressed the United Nations, the World Economic Forum and several parliaments declaring: "Our house is on fire". She was named Time magazine's Person of the Year in 2019 and was nominated for the Nobel Peace Prize.

Concern about climate change was simultaneously growing within the establishment. The International Panel on Climate Change delivered its starkest warning yet about the planet heating up saying that greenhouse gas emissions were fast approaching an irreversible tipping point. Meanwhile extreme weather events and huge wildfires – 36,000 in the Amazon rainforest alone - grabbed headlines, with July 2019 being officially declared the hottest month on record. Two months later there were two climate marches which attracted 6 million and 4 million people respectively worldwide, and a funeral service was held in Iceland for a glacier that had melted.

There were huge protests on every continent during these two years about issues ranging from gun violence to immigration policies, government corruption and Brexit, some of which became extremely violent with loss of life.

Perhaps most prominent was the student uprising in Hong Kong, the largest series of demonstrations the territory had ever seen, with more than 1 million people taking part in rallies protesting extradition laws. The protests became increasingly violent and confrontational over the next year.

We can correlate the Hong Kong protest movement with Comet Wirtanen's alignment with Uranus at 29 Aries, very close to the territory's Ascendant in its 1997 inception chart. When the comet reached perihelion

Hong Kong 1997
Natal
1 Jul 1997, Tue
01:30 AWST -8:00
Hong Kong, China
Geocentric
Tropical
Placidus

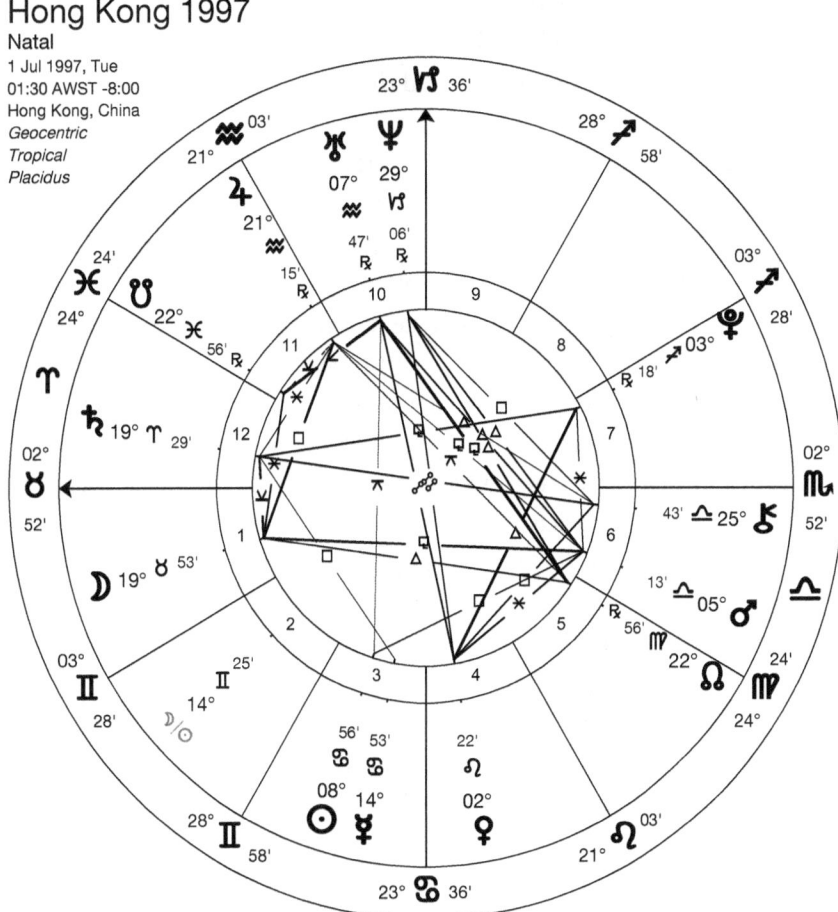

Figure 8.12.2

at 20 Taurus on 12 December 2018 it was conjunct Hong Kong's Moon. (Figure 8.12.2)

Covid-19

When the comet became visible to the naked eye it was in the constellation Fornax in the southern sky, named by French astronomer Nicolas Louis de Lacaille in 1756 after the symbol of the chemical furnace.

One year after Comet Wirtanen appeared, the Covid 19 pandemic began in Wuhan, China, leading to the death of many millions around the world, locking down countries and dealing massive blows to local economies as small businesses closed and people were forced to stay at home. The pandemic and its global impact have clear associations with the powerful Jupiter-Saturn-Pluto conjunction in Capricorn in January 2020, after which many astrologers predicted the world would never be the same again.

We can also relate the pandemic to the comet's discovery and passage though the Fornax constellation with the source of the SARS-CoV-2 virus being widely recognised as US-funded "gain of function" research being carried out with bat viruses in a laboratory in Wuhan, China.

We can also see the Fornax constellation reflected in the major role the pharmaceutical industry played with the search for effective vaccines, unprecedented vaccine roll outs, new mRNA vaccine technology being used at scale for the first time after much shorter vaccine trial periods than normal, all of which sparked huge controversy as alternative treatments were suppressed and criticism of these policies "cancelled".

With the comet's journey through the constellation Auriga, the chariot, I predicted transport restrictions, though I never suspected the scale introduced by the lockdown. I also saw the alignment with fixed star Capella raising issues around freedom of movement.

The comet's passage through the Taurus constellation also reflected the pandemic's huge economic impact, including the largest transfer of wealth the world has ever seen, the unprecedented level of government support payments and massive shocks to financial markets.

Looking at Fornax again, a chemical factory in China exploded in March 2019 killing at least 78 people and injuring more than 600. It was so powerful, it initially registered as an earthquake. And in Russia there was a gas explosion in a biochemical plant storing Ebola, smallpox and anthrax viruses.

Other events

The comet also travelled through the constellation Cetus, the sea monster, and this period saw Japan recommence whaling after leaving the International Whaling Commission.

It also travelled through Eridanus, the flowing river, and flooding caused many deaths, injuries and financial losses in China, the Philippines, India, Somalia and Kenya, where tens of thousands were left homeless. A dam collapse in Laos left 1,100 people missing, and the whole world held its breath while rescuers spent days trying to reach 12 teenagers and their football coach trapped in a cave in Thailand by floods.

Finally the comet's passage through the highly virile constellation Perseus may relate to the sentencing of film producer Harvey Weinstein to 23 years in jail for sexual assault in January 2020 following his arrest in 2018, which sparked the global #MeToo social media movement highlighting the scale of sexual abuse women suffer throughout the world.

Comet Neowise (C/2020 F3)

Type of comet: long period

Orbital period: 4,500-6,800 yrs

Astronomical data

	Date	Position
Discovery	27 Mar 2020	13 Leo
First seen	20 Jun 2020	29 Gemini
Last Perihelion	03 Jul 2020	29 Gemini
Last seen	Mid Aug 2020	20-25 Libra
Crossed celestial equator	03 Jun 2020 17:33 UT	01 Cancer
Crossed ecliptic	29 Jun 2020 01:57 UT	29 Gemini

Tropical zodiac

Date	Sign
20 Jun 2020 - 03 Jul 2020	Gemini
03 Jul 2020 - 18 Jul 2020	Cancer
18 Jul 2020 - 25 Jul 2020	Leo
25 July 2020 - 02 Aug 2020	Virgo
02 Aug 2020 - 23 Aug 2020	Libra

Constellations

Note: not to be confused with the 12 signs of the tropical zodiac.

Date	Constellation
20 Jun 2020 - 28 Jun 2020	Orion
28 Jun 2020 - 02 Jul 2020	Taurus
02 Jul 2020 - 12 Jul 2020	Auriga
12 Jul 2020 - 17 Jul 2020	Lynx
17 Jul 2020 - 29 Jul 2020	Ursa Major
29 Jul 2020 - 09 Aug 2020	Coma Berenices
09 Aug 2020 - 12 Aug 2020	Virgo
12 Aug 2020 - 15 Aug 2020	Bootes
15 Aug 2020 - 30 Aug 2020	Virgo

Planetary alignments

Date	Planet	Zodiac location
23 Jun 2020	Moon's North Node	29 Gemini
09 Jul 2020	Mercury Rx	5 Cancer
16 Jul 2020	Sun	24 Cancer

Several comets appeared in the sky during 2020, but none attracted attention like Comet Neowise, the brightest comet since Hale-Bopp in 1997, affording it the title Great Comet of 2020.

Comet Atlas had been expected to make a dramatic appearance in March, but it broke up before becoming visible. Comet Swan could be seen by the naked eye, but only for a few days as it travelled between constellations Cetus and Pisces.

Comet Neowise was first detected in March but could only be observed in the morning sky once it appeared far enough away from the Sun so as not to be outshone. When first seen on 20 June just before sunrise it was in the constellation Orion the hunter.

Events

Covid pandemic

The first year of the 2020s changed the world. By the end of the year the SARS-CoV-2 virus was being held responsible for more than 80 million cases of respiratory disease worldwide and more than 1.5 million deaths. Most of the world had been placed under lockdown for nine months or more, and mass vaccination programmes were in play. Quite correctly astrologers focused their attention on the Saturn-Pluto conjunction at 22 Capricorn on 12 January 2020 with the Sun and Mercury within a degree on either side.

But shortly afterwards Comet Neowise was discovered by astronomers in March just as lockdowns were being imposed, restricting people to their homes and their immediate neighbourhood. When it was first seen from Earth in June it was flying through the tropical zodiac sign Gemini, the sign of transport, communication and gossip.

The year was full of conflicting stories about the source of the pandemic and the appropriate way of tackling it. "Conspiracy theories" thrived, individuals including doctors and academics were censored if they expressed doubts about official policy, arguments and dissent about the lockdowns and the vaccines exploded. There were scandals about politicians and civil servants breaching the rules they were imposing on everyone else, while corrupt business deals took advantage of the crisis.

Our way of interacting with each other was transformed. With almost everyone working from home, video conferencing technologies like Zoom and Microsoft Teams came into their own, and people suddenly had to spend more time with their families. Here we can see a reflection in the comet's passage through Cancer where the comet was at its brightest. People were having to pay more attention to their home environment, many deciding to leave the metropolis and move to the relative safety of the countryside to protect themselves and their loved ones.

The comet then moved through Leo, a sign concerned with personal sovereignty and pleasure seeking. The pandemic and the vaccine mandates felt to many like an invasion of their right to make decisions for themselves. The censorship denied people's right to even express a point of view. People lost their jobs for taking a stand against the prevailing narrative.

Meanwhile the Leo-ruled entertainment industry was completely transformed as people were denied the ability to go out to bars or restaurants, the cinema or any public event. As a result online streaming, gaming and social media filled the vacuum, while many used the opportunity to spend more time outside going for walks, enjoying the fresh air and being in nature.

And, of course, the constellation Auriga, the chariot, through which the comet travelled relates to freedom of movement, which was so severely restricted.

The US election – Trump v Biden

In the middle of the election campaign, the comet was discovered at 13 Leo, close to sitting president Donald Trump's Pluto. After it became visible it crossed Trump's natal Mercury, Saturn and Venus in Cancer. On 9 July it aligned with retrograde Mercury at 5 Cancer, close to his natal Mercury. On 16 July it aligned with the Sun at 24 Cancer, the midpoint of his tight conjunction of natal Saturn and Venus, his Taurus MC ruler. Finally, before disappearing from view in late August, it transited Trump's natal Pluto, Mars and Ascendant. (Figure 8.13.1)

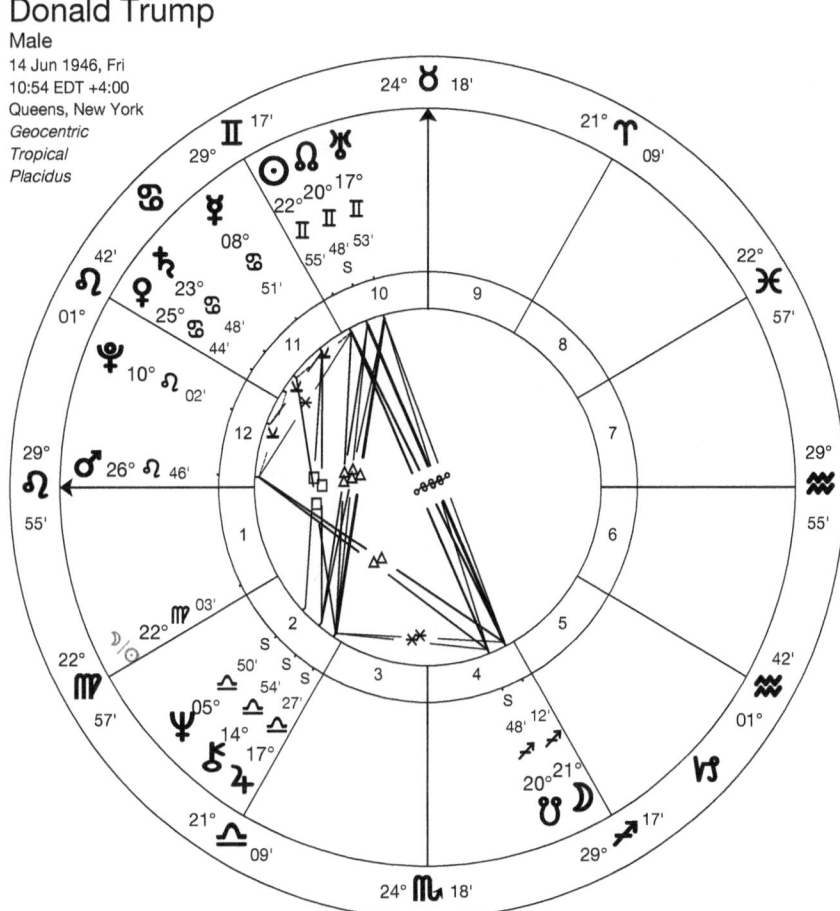

Figure 8.13.1

Comet Neowise had less impact on the successful Democrat candidate Joe Biden's chart, failing to align with any of his personal planets. Its alignment with the Sun at 24 Cancer was within a degree of Biden's Jupiter, and before it disappeared it transited his natal Pluto, Chiron and the North Node. (Figure 8.13.2)

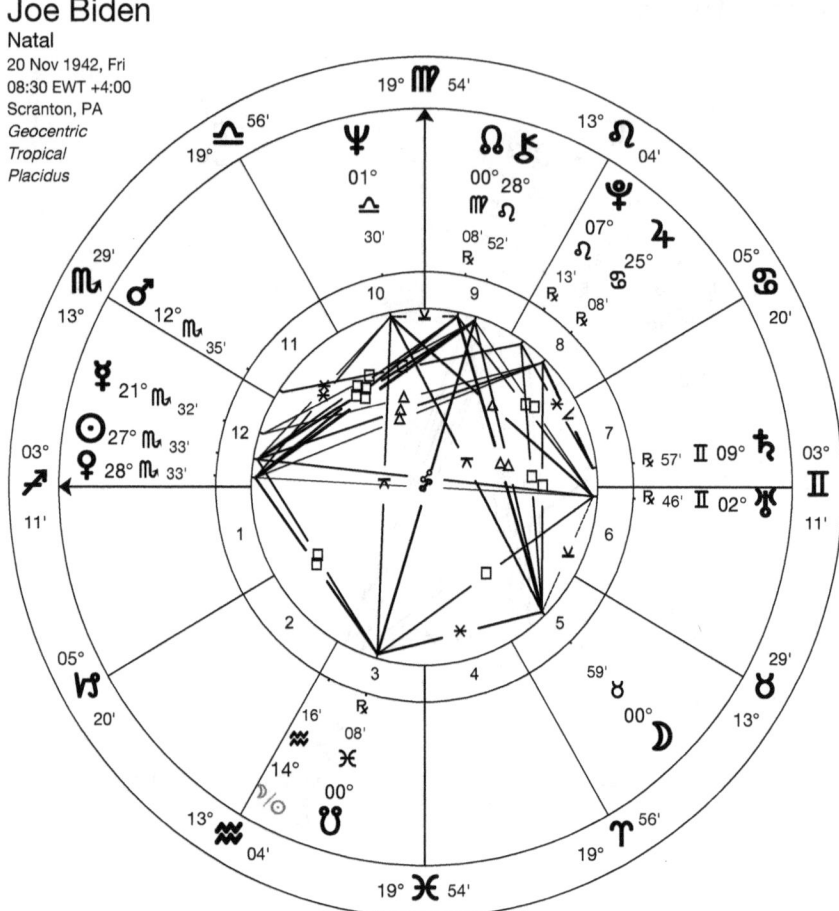

Figure 8.13.2

Trump's refusal to accept defeat and the legitimacy of the election result led to rioters invading the US Capitol on 6 January 2021 to prevent Biden's victory from being formalised by Congress, raising nationwide fears that US democracy was under threat. Comet Neowise certainly impacted the US Sibly chart. Its alignment with retrograde Mercury at 5 Cancer was right on the chart's Jupiter, while its alignment with the Sun at 24 Cancer was on the chart's retrograde Mercury in the 8th house. (Figure 8.13.3)

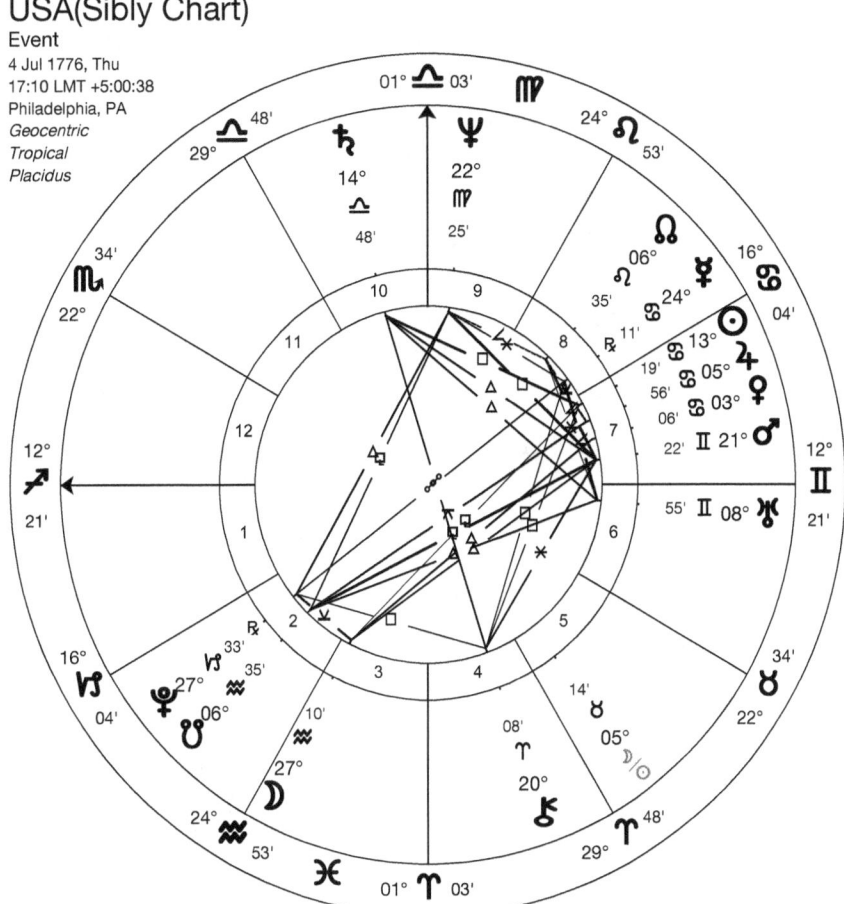

Figure 8.13.3

The GameStop short squeeze

In 2020 the video game chain store GameStop Corp was in trouble, closing stores and watching its share price fall. A number of hedge funds attempted to profit from the company's decline by instigating a "short squeeze", essentially betting on its downfall. This move by professional investors was publicised by popular YouTuber Keith Patrick Gill whose channel *Roaring Kitty* has more than 1 million subscribers. Gill successfully urged his followers to fight back against the hedge funds by purchasing GameStop shares, pushing the share price up from around $40 to more than $500. The short squeeze ended up costing the professionals more than $5 billion.

The bankers fought back by putting pressure on the online investor platform Robinhood, which had been processing the share purchases by the small-scale investors who were following *Roaring Kitty*. Robinhood buckled under the pressure and started to restrict transactions, a move that raised concerns about Wall Street's ability to manipulate the market and even prompted a congressional investigation.

The incident is highly significant in that it demonstrates a shift in power, if only a small one, within the investment industry. Thanks to the growth of social media investment platforms, small scale "retail investors" can now play the markets. While the professional investors ridicule them as "dumb money", the GameStop incident reflects the power of the crowd and could be viewed as an uprising against the top heavy financial system.

Here we can see the influence of Gemini, which governs the rapid flow of information on which the investors rely, also Cancer's concern with protecting the weak and vulnerable, and even Leo, which connects with sovereignty and doesn't like to be pushed around. You could even bring in the constellations Taurus, the Bull, and Ursa Major, the Great Bear, symbolising the different approaches the large and small investors took.

Two key individuals involved in this episode deserve our attention astrologically in connection with Comet Neowise:

Keith Patrick Gill, who led the retail investor uprising through his YouTube channel *Roaring Kitty*, has natal Mercury at 5 Cancer, the exact degree Comet Neowise aligned with retrograde Mercury in the sky on 9 July. His natal Venus sits at 21 Cancer, a degree the comet crossed on its way to align with the Sun at 24 Cancer. The symbolism is clear with Mercury, the planet of communication, and Venus, the planet governing wealth and value, triggered by the comet in Cancer, the sign that wants to protect the weak against the strong. (Figure 8.13.4)

Andrew Left, an activist short seller who writes for and edits the online investment newsletter *Citron Research*, lost a fortune when the short squeeze boomeranged on him and other investors hoping to profit from GameStop's collapse. As a result of the incident, Left abandoned his long-held practice of short selling, betting on companies failing, and later claimed he and his family had been victims of an online harassment campaign. In 2024 he was arrested and accused of misleading investors. Left's birth chart has a stellium in Cancer with the Sun, Mercury and Mars,

Comet Case Studies 147

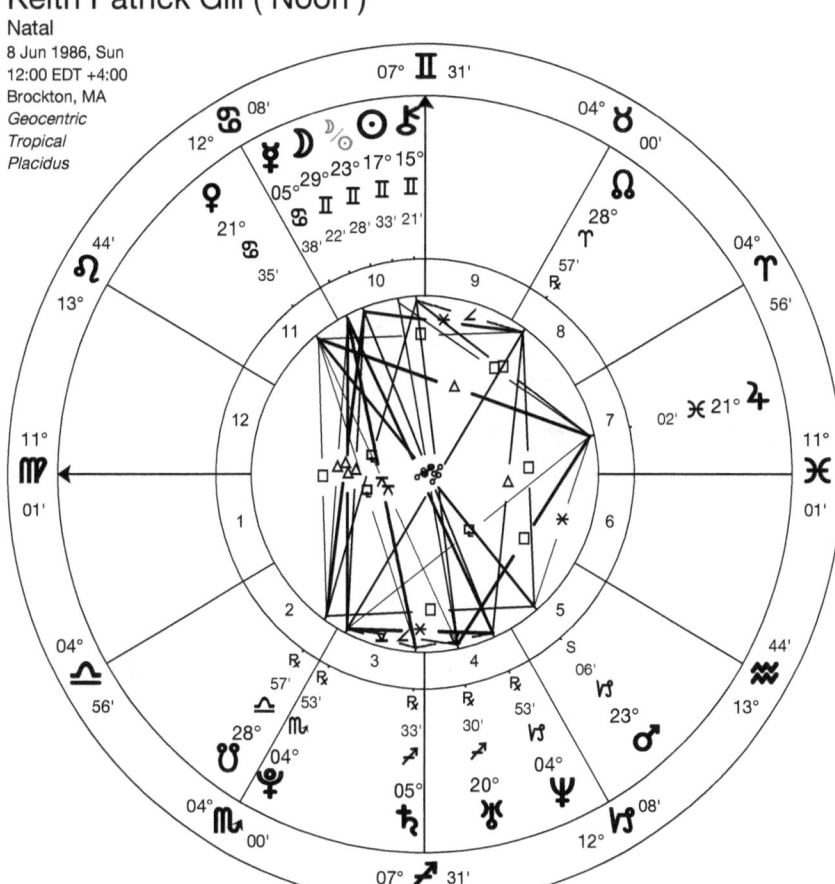

Figure 8.13.4

and Venus at 26 Leo. The comet flew past all these natal planet positions, aligning with the Sun at 24 Cancer on 16 July right on natal Mars.

The GameStop incident has subsequently been turned into a movie called *Dumb Money*, based on the book called *The Antisocial Network* by Ben Mezrich.

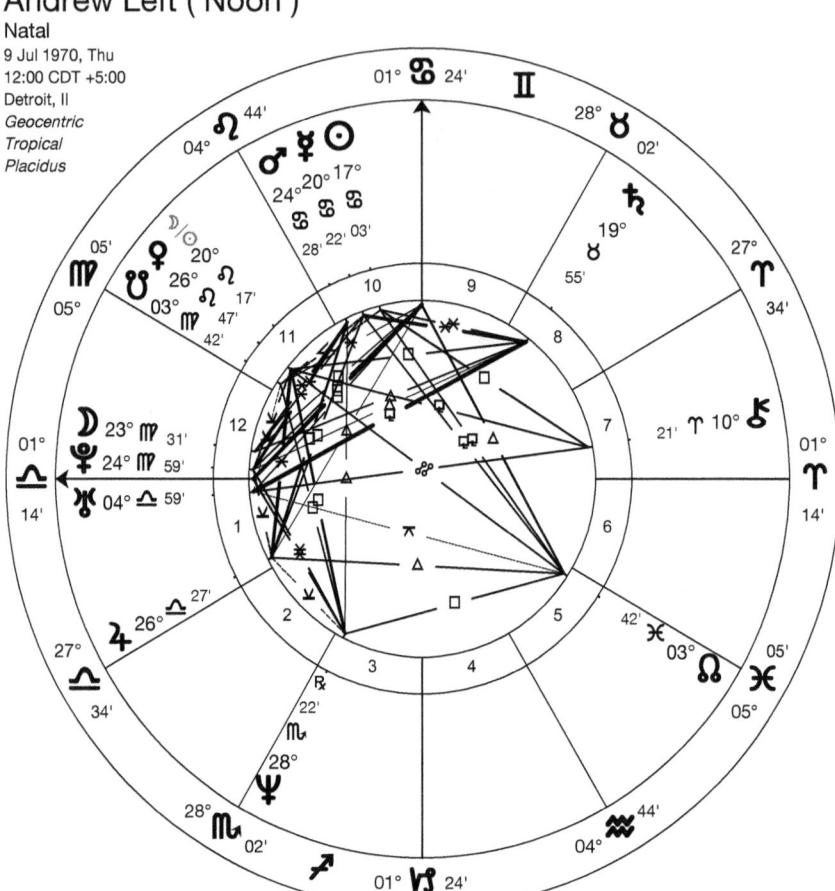

Figure 8.13.5

Comet Neowise through the Elements

We have all the elements in play with this comet passing through air with Gemini and Libra, water with Cancer, fire with Leo and earth with Virgo.

Air

2020 saw record breaking hurricanes, cyclones and storms. In May Cyclone Amphan killed 100 people in India and Bangladesh, but 4 million people had to be evacuated and damage worth $13 billion made it the costliest cyclone in history. In the US a massive thunderstorm also broke the record of causing the most damage when it devastated much of the country's mid-west, especially Nebraska, Iowa, Illinois, Wisconsin and

Indiana. Hundreds died in hurricanes in Nicaragua, Vietnam, Alabama and Louisiana, while the Philippines experienced Typhoon Goni, the most powerful landfalling tropical cyclone in history, killing dozens and displacing hundreds of thousands of people.

Water

Hundreds of lives were lost in massive floods that affected Indonesia, Somalia, Vietnam, Japan, South Korea, India and Nepal where 4 million people were left homeless when the Brahmaputra river flooded, causing 189 deaths.

There were also two terrible oil spills, including the second largest in Russian history when 20,000 tonnes of oil leaked into the Ambarnaya river in Siberia, and the island of Mauritius in the Indian Ocean experienced its worst ever environmental disaster when Japanese bulk carrier *Wakashio* broke in half after stranding on a reef and spilled 1,000 tonnes of oil into the surrounding sea. Both incidents were declared national emergencies. The deadliest shipwreck of the year saw 140 migrants drown off the coast of Senegal on their way to the Spanish Canary Islands.

Fire

There was no shortage of deadly military skirmishes during the year, though perhaps the number was reduced thanks to the pandemic. The year began with the shocking assassination of Iran's top general Qasem Soleimani and Iraqi paramilitary leader Abu Mahdi al-Muhandis by a US drone in Iraq and ended with the assassination of Iran's chief nuclear scientist Mosen Fakhrizadeh. Beirut experienced a colossal explosion when a warehouse full of poorly stored ammonium nitrate blew up, killing 218 people, injuring thousands, leaving 300,000 homeless and causing damage estimated to be worth up to $15 billion.

Despite the worldwide lockdown, 2020 turned into a year of major protests as people got fired up. It started in May with the murder of George Floyd in Minneapolis by white policeman Derek Chauvin sparking the Black Lives Matter protests that spread across the entire world. In eastern Europe there were huge anti-government protests in Romania, Bulgaria and Serbia and the biggest protest in the history of Belarus followed the fifth election victory of President Alexander Lukashenko. Meanwhile East Asia saw China crack down on the Hong Kong protests after introducing new laws and jailing the movement's leaders, while the government in Thailand also suppressed demonstrations and imposed media censorship after declaring a "severe" state of emergency.

Earth

The year saw major earthquakes in Croatia and the Aegean Sea, while 30 people died in a landslide in Vietnam. A Philippine volcano erupted for the first time in more than 40 years.

The pandemic had massive impacts on the world economy, which virtually shut down for the whole year. 2020 saw the stock market's worst week since the Global Financial Crisis of 2008 and the Dow Jones Industrial Average experienced it largest drop since Black Monday of 1929. While small businesses were bailed out by massive government handouts, vast wealth was transferred to some of the richest men in the world, especially the tech giants. Amazon founder Jeff Bezos became the first person to attain a net worth of $200 billion. Meanwhile documents leaked to the International Consortium of Investigative Journalists exposed $2 trillion of suspect transactions across global financial institutions.

Bones found in Bulgaria revealed that *homo sapiens* arrived in Europe much earlier than previously thought, changing perceptions of our ancestry. Researchers announced the discovery of the world's oldest-known land animal, the fossil of a Scottish millipede that lived 425 million years ago. And the largest ever cache of 200 mammoth skeletons was discovered during the construction of a new airport in Mexico.

The comet was last seen in the air sign of Libra, a sign connected with diplomacy. Despite the violence in certain parts of the world, 2020 did see a number of peace agreements between nations:

- The US signed a conditional peace agreement with the Taliban in Afghanistan and started the process of withdrawing troops from the country;
- Saudi Arabia declared a ceasefire with the Houthis of Yemen;
- Israel signed peace accords with Sudan, the United Arab Emirates, Bahrain, and Bhutan;
- Serbia and Kosovo signed a peace agreement;

- the warring sides in Sudan did the same; and
- while fighting broke out between Armenia and Azerbaijan, the two sides agreed a ceasefire within six weeks.

Fifteen countries in the Asia-Pacific region signed the Regional Comprehensive Economic Partnership to become the largest free trade bloc in the world, representing one third of the world's population. In contrast, the UK left the European Union and the US pulled out of the Paris Agreement on tackling climate change.

Comet Leonard C/2021 A1

Comet type: long period (inbound); hyperbolic (outbound)

Orbital period: 80,000 years (inbound)

Astronomical data

	Date	Position
Discovery	03 Jan 2021	08 Libra
First seen	05 Dec 2021	18 Libra
Perihelion	03 Jan 2022	15 Aquarius
Last seen	11 Jan 2022	15 Aquarius
Crossed celestial equator	12 Dec 2021 06:20 UT	13 Sagittarius
Crossed ecliptic	16 Dec 2021 21:14 UT	17 Capricorn

Tropical zodiac

Date	Sign
05 Dec 2021 - 07 Dec 2021	Libra
07 Dec 2021 - 10 Dec 2021	Scorpio
10 Dec 2021 - 14 Dec 2021	Sagittarius
14 Dec 2021 - 20 Dec 2021	Capricorn
20 Dec 2021 - 11 Jan 2022	Aquarius

Constellations

Note: not to be confused with the 12 signs of the tropical zodiac.

Date	Constellation
05 Dec 2021 - 08 Dec 2021	Bootes
08 Dec 2021 - 10 Dec 2021	Serpens
10 Dec 2021 - 14 Dec 2021	Ophiuchus
14 Dec 2021 - 15 Dec 2021	Scutum
15 Dec 2021 - 20 Dec 2021	Sagittarius
20 Dec 2021 - 29 Dec 2021	Microscopium
29 Dec 2021 - 11 Jan 2022	Piscis Austrinus

Fixed Stars

Date	Star
13 Dec 2021	Sinistra (Nu Ophiuchi)
16 Dec 2021	Albaldah (Pi Sagittarii)

Planetary alignments

Date	Planet	Zodiac location
10 Dec 2021	Mars	27 Scorpio
13 Dec 2021	Sun	21 Sagittarius
15 Dec 2021	Venus	06 Capricorn
16 Dec 2021	Mercury	13 Capricorn
19 Dec 2021	Pluto	27 Capricorn

Comet Leonard was the brightest comet of 2021 when it became visible at the end of the year in the sign of Libra on its 80,000 year orbit. It travelled rapidly across the sky passing through five zodiac signs before disappearing, never to be seen again as astronomers calculate it will break free from our solar system on its outward journey, which is what is meant by the term 'hyperbolic".

Astronomers predicted it would be very bright, but it proved inconsistent with three separate outbursts in December, taking it to magnitude 3 before fading back to magnitude 4. Its tails were described as "some of the best ever observed". It also came closer than any other comet in recorded history to one of our solar system's planets, when it reached 4.26 million kilometres from Venus. Skywatchers saw it start to disintegrate as it headed back out into space in March 2022.

Events

Russia v Ukraine

The build up to the Ukraine conflict and the launch of Russia's military operation in February 2022 took place during the period Comet Leonard was visible in the sky.

Astrologer Nick Campion lists two separate independence charts for Ukraine in *The Book of World Horoscopes*. The most recent one is based on the country's 1991 independence referendum and reveals Comet Leonard would have passed every planet in Ukraine's chart bar Jupiter, and also Chiron. (Figure 18.14.1)

The comet was discovered at 8 Libra, very close to Ukraine's IC, and became visible at 18 Libra less than a degree from its 4th house Moon, both of which represent the homeland. Over the course of its five week period of visibility, it aligned with Venus in Libra, Pluto in Scorpio,

Mars-Sun-Mercury in Sagittarius, North Node-Uranus-Neptune in Capricorn and Saturn in Aquarius.

On its journey Comet Leonard aligned with transiting Sun at 21 Sagittarius close to Ukraine's Mercury, and then with transiting Mercury at 13 Capricorn very close to the country's North Node-Uranus-Neptune conjunction.

It is worth noting that Ukraine has the same outer planetary positions as the rest of the eastern European countries that declared independence after the collapse of the Soviet Union.

Comet Leonard passed by the star Albadah in the Sagittarius constellation, which is on the ecliptic at around 16 Capricorn. The star's name means 'city', or 'town', and I have observed that when outer planets cross

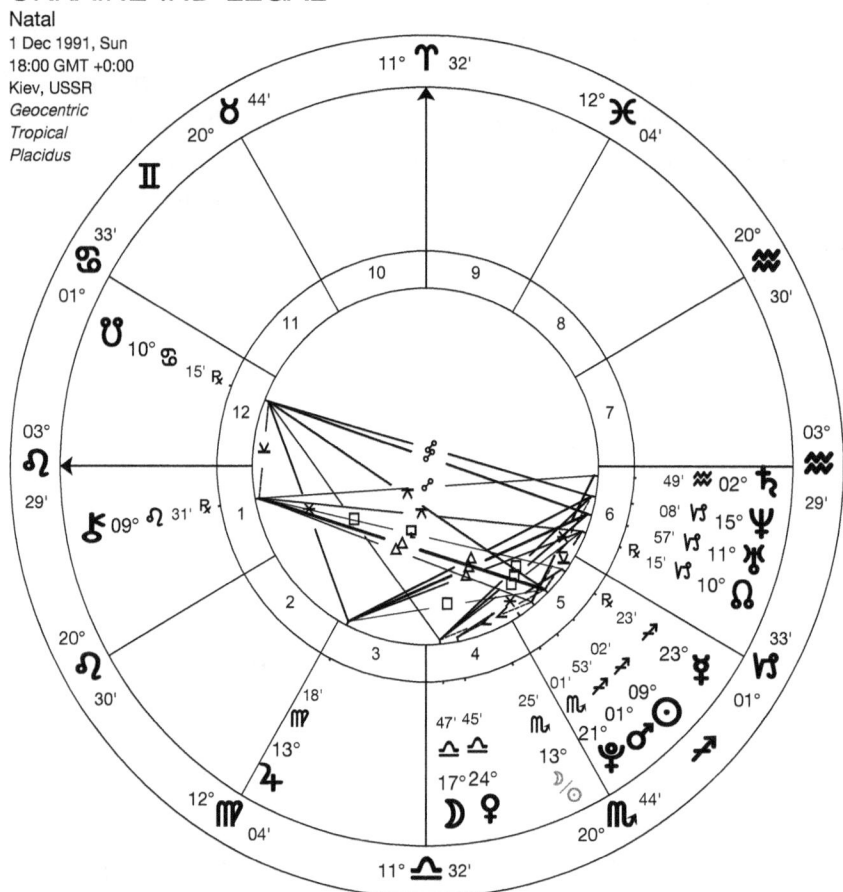

Figure 8.14.1

this point we often see war and civil unrest. A good example is Neptune's alignment with Albadah when the Soviet Union collapsed.

United Kingdom

2022 proved to be a momentous and extremely destabilising year for Great Britain with the death of its longest serving monarch and two prime ministers resigning.

Queen Elizabeth II's death marked the end of an era for Britain and the Commonwealth over which she had "ruled".

Comet Leonard can be seen as a harbinger of her passing, aligning with transiting Mars at 27 Scorpio on 10 December 2021 right on the Queen's MC, her public profile, and close to her ruling planet Saturn. (Figure 8.14.2)

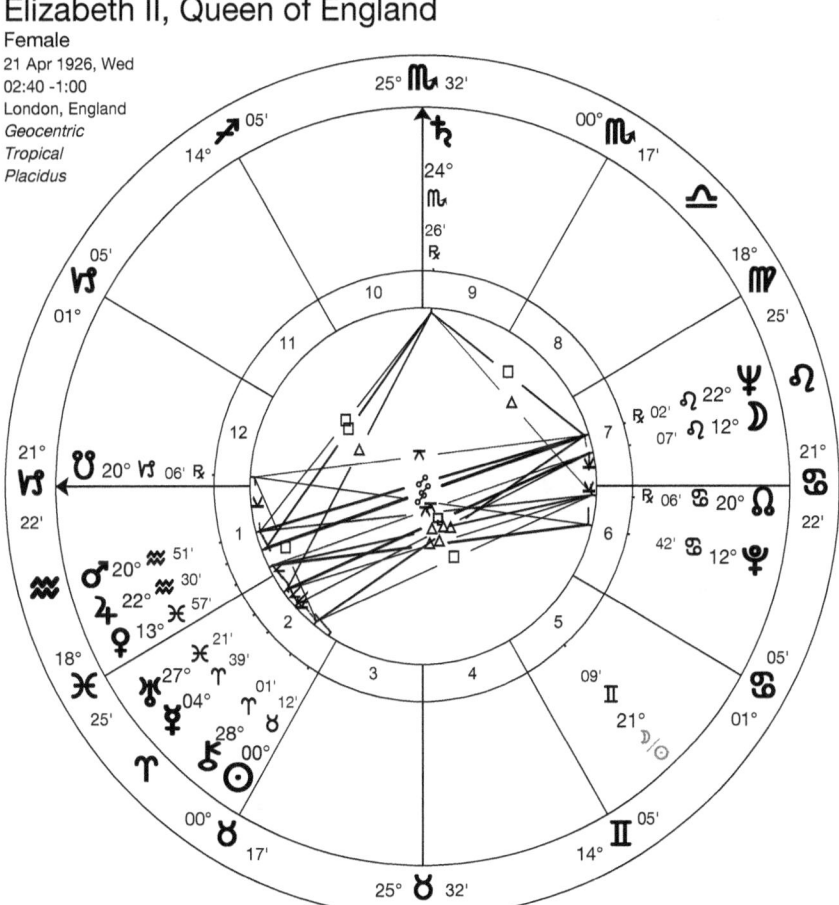

Figure 8.14.2

She was succeeded by her eldest son Charles, whose natal Venus at 14 Libra and Neptune at 16 Libra lie close to the point where the comet was first seen with the naked eye, at 18 Libra. It then flew past his natal South Node, Mercury and Sun before aligning with Mars close to natal Chiron. On 13 December 2021 it aligned with the transiting Sun within minutes of his natal Mars, ruler of the new king's MC in Aries. (Figure 8.14.3)

Prime minister Boris Johnson was forced out when one third of his government resigned after he appointed Chris Pincher MP as deputy chief whip, a party enforcer, despite knowing of sexual assault allegations against him.

On 13 December 2021 Comet Leonard aligned with the fixed star Sinistra (Nu Ophiuchi), which, according to astrologer Vivian Robson, has

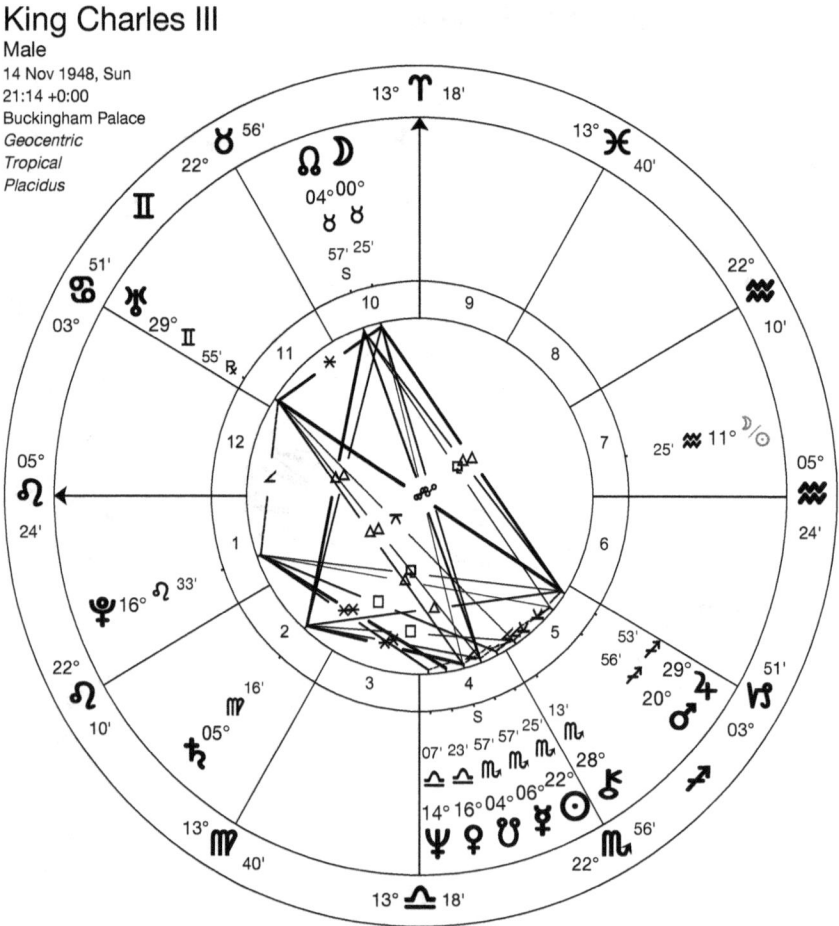

Figure 8.14.3

an 'immoral, mean and slovenly nature'. As it is not on the ecliptic, it is rare for this star to be highlighted in astrology and represents a clear indication of scandal. Its position aligns with 28 Sagittarius on the ecliptic, exactly opposite Johnson's natal Venus and Sun. Also the comet was first observed by the naked eye at 8 Libra close to his Ascendant, another sign of trouble for the hapless PM. (Figure 8.14.4)

Pincher himself has a natal Uranus-Mercury-Jupiter conjunction close to 8 Libra. His natal Neptune at 26 Scorpio is one degree from the comet's alignment with transiting Mars.

Johnson was replaced by Liz Truss, who became Britain's shortest-serving prime minister, stepping down after less than two months in office after introducing extreme budgetary measures that triggered a huge

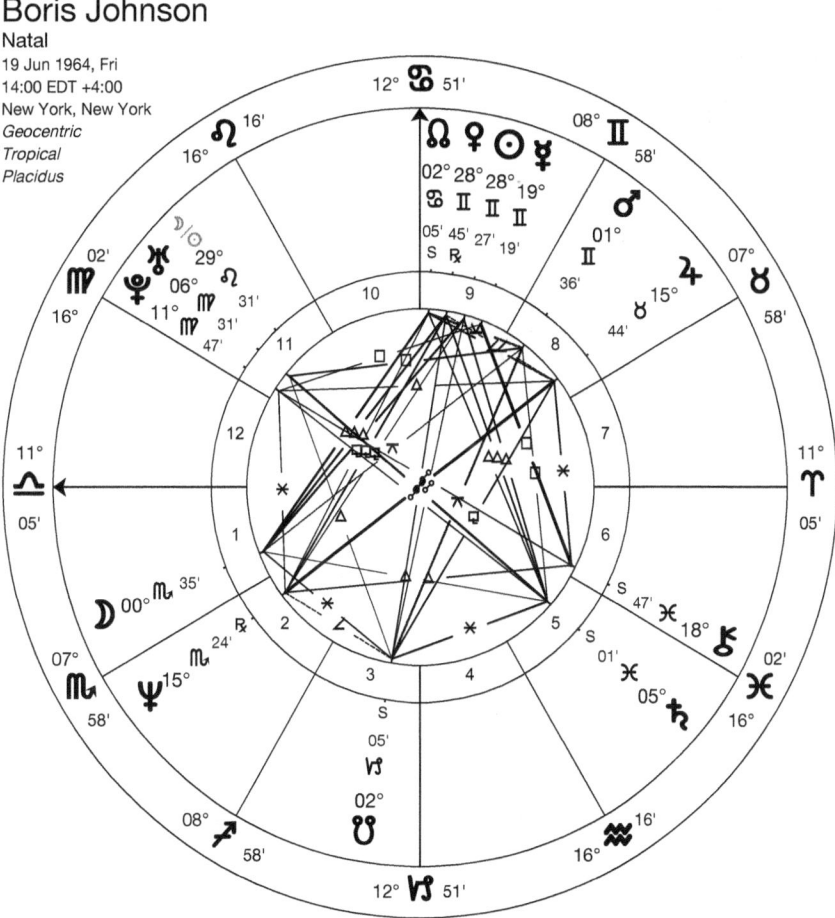

Figure 8.14.4

158   *Comets in Astrology*

backlash and took the value of the pound sterling to a record low. She resigned after being polled as the most unpopular prime minister in British history and was replaced by multi-millionaire Rishi Sunak, Britain's first non-white head of government.

Britain's astrological chart of 1801 (Figure 8.14.5) has an Ascendant at 7 Libra, just one degree from where Comet Leonard was discovered and square the chart's Sun at 10 Capricorn, symbolising the country's leadership. While visible, the comet passed the chart's Neptune, Chiron, Mercury and Sun before reaching perihelion and disappearing at 15 Aquarius, beside the UK's Venus.

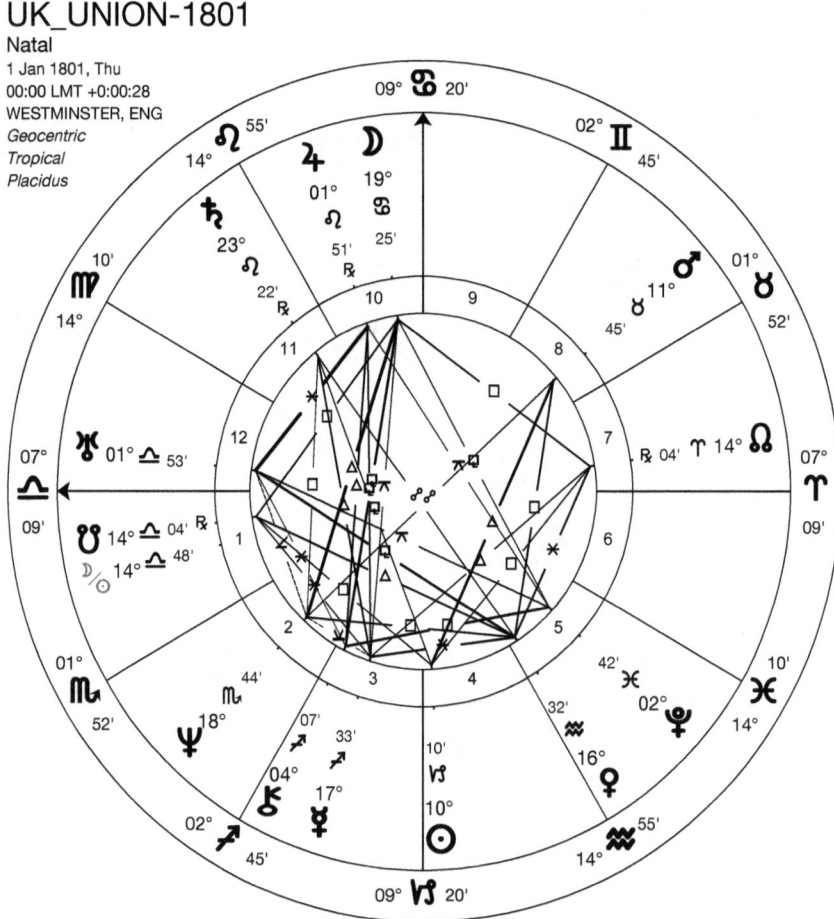

Figure 8.14.5

Technology – AI, Twitter & Musk

New technologies, especially in computing, often experience breakthroughs within a year of a comet's appearance.

In 2022 artificial intelligence (AI) broke through into mass public awareness when local artist Jason Allen won first prize in an art competition at the Colorado State Fair in August with an AI-generated image, stirring up a fierce debate around copyright and what constitutes art.

On 30 November research organisation OpenAI, based in San Francisco, released the generative AI chatbot ChatGPT. It rapidly became the fastest growing software application in history gaining 102 million

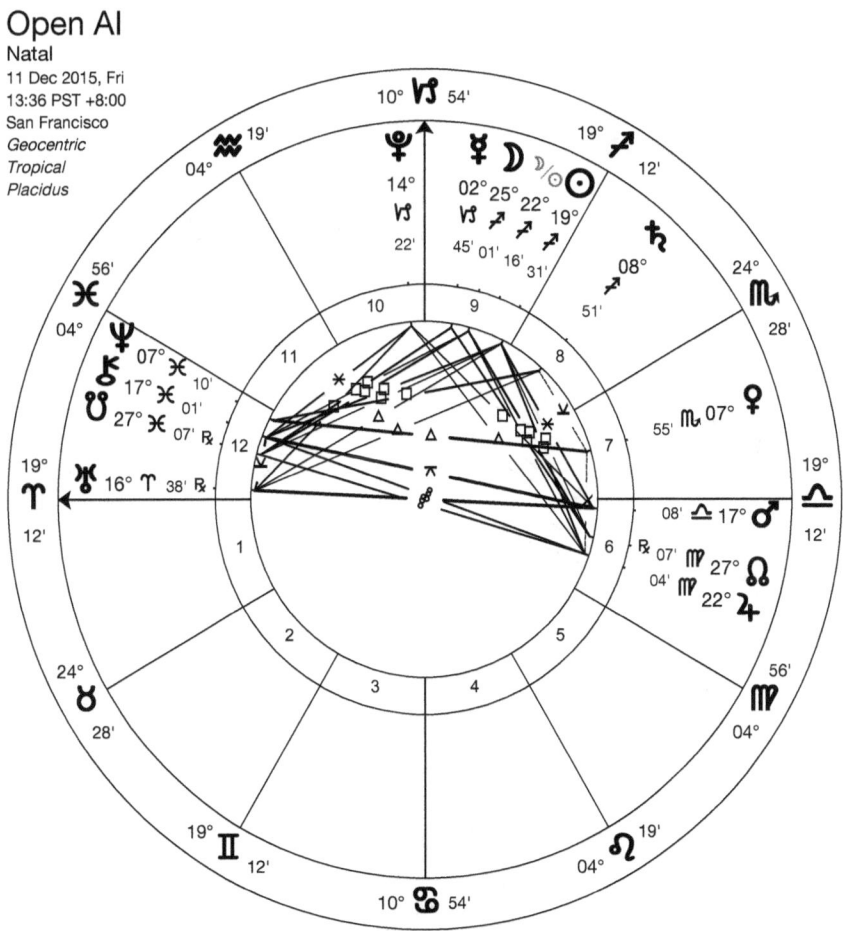

Figure 8.14.6

users in just two months to become one of the top 10 most-visited websites globally.

OpenAI was founded as a non-profit organisation in San Francisco at 13:36 PST on 11 December 2015. The chart ruler is Mars at 17 Libra, one degree from where Comet Leonard was first seen. On 13 December it aligned with the Sun transiting at 21 Sagittarius, just two degrees from the Sun in OpenAI's chart, and flew past the chart's Moon at 25 Sagittarius before aligning with transiting Mercury at 13 Capricorn, one degree from where OpenAI's Pluto is positioned. (Figure 8.14.6)

The chart for ChatGPT's release at 10:02 PST on 30 November 2022 has an Ascendant at 17 Capricorn, the exact degree where Comet Leonard crossed the ecliptic on 16 December 2021.

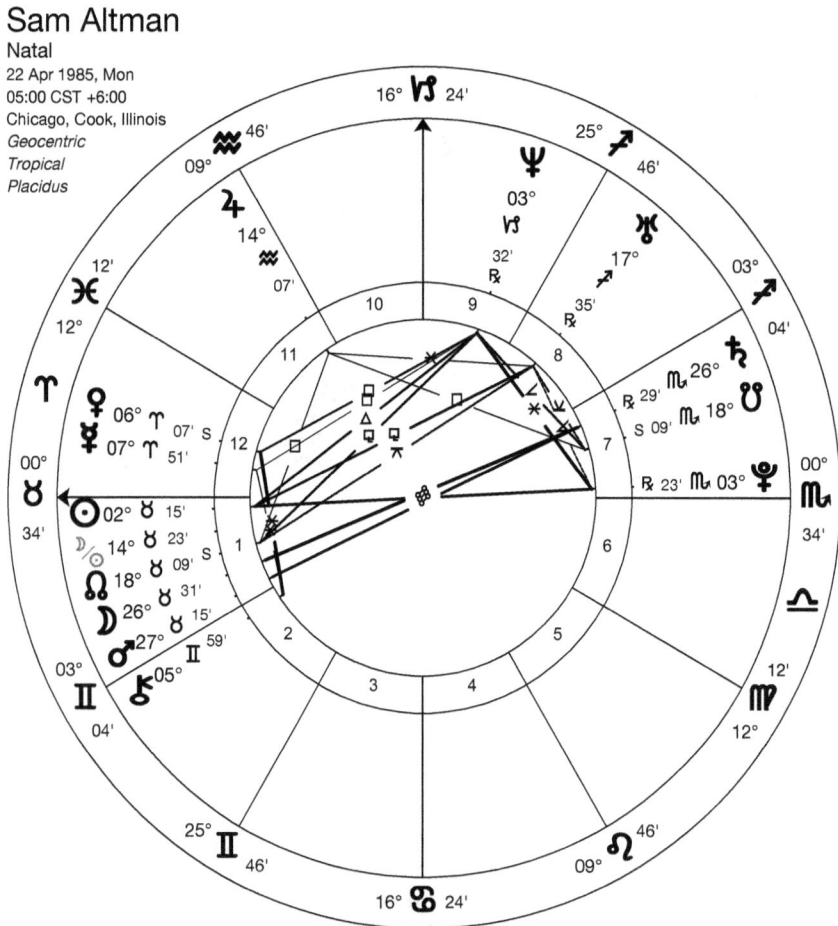

Figure 8.14.7

OpenAI's CEO Sam Altman's birth chart has Jupiter at 14 Aquarius, one degree from the comet's perihelion and the point at which it disappeared. His MC is at 16 Capricorn, where the comet crossed the ecliptic, and his natal Saturn at 26 Scorpio is one degree from where the comet passed transiting Mars. (Figure 8.14.7)

Tech giant and "world's richest man" Elon Musk was a founding partner in OpenAI but left the board of directors in 2018.

In April 2022, three months after Comet Leonard disappeared from view, Musk completed his takeover of social media company Twitter, renaming it X.

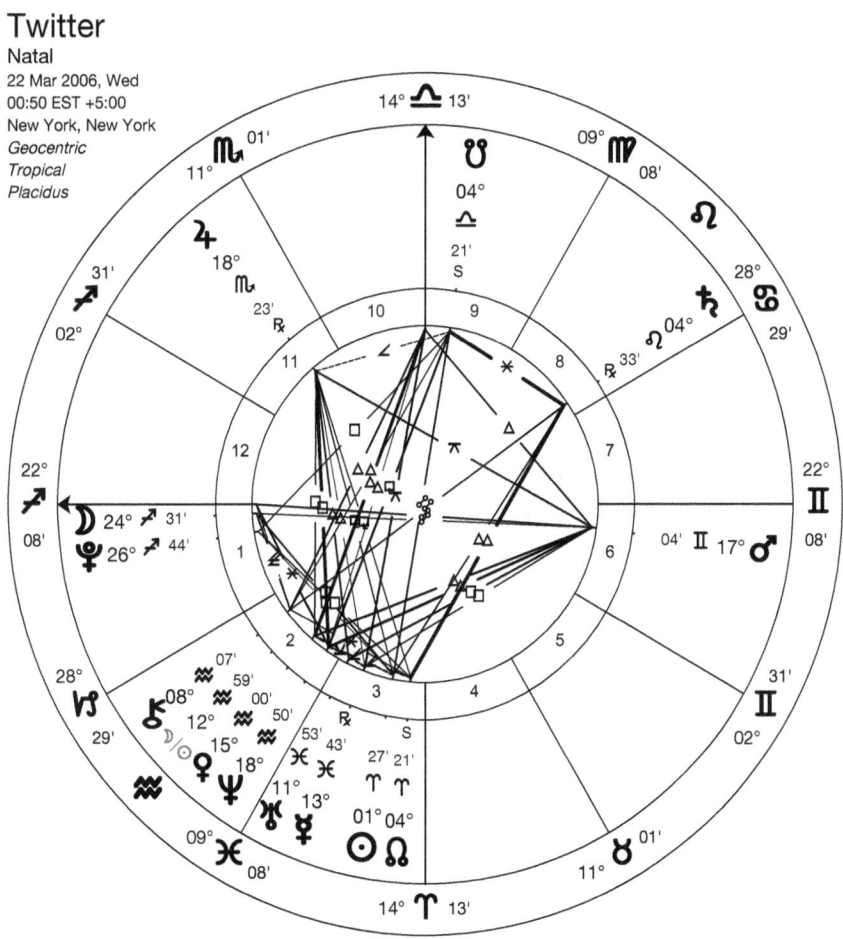

Figure 8.14.8

Musk's natal Venus is just two degrees ahead of Mars in the foundation chart of Twitter, both in the sign of chat and gossip, Gemini. Comet Leonard's appearance will have triggered both.

Twitter was founded on 22 March 2006 with an Ascendant at 22 Sagittarius and the Moon at 24 Sagittarius, very close to where the comet aligned with the transiting Sun at 21 Sagittarius. It appeared at 18 Libra close to the MC and it reached perihelion at 15 Aquarius right on Twitter's Venus and conjunct the channel's Neptune at 18 Aquarius. All of this suggests its arrival would imply important changes for the social media platform. (Figure 8.14.8)

Musk's North Node is at that perihelion degree of 15 Aquarius, the comet was discovered at 8 Libra close to his natal Uranus, and it aligned

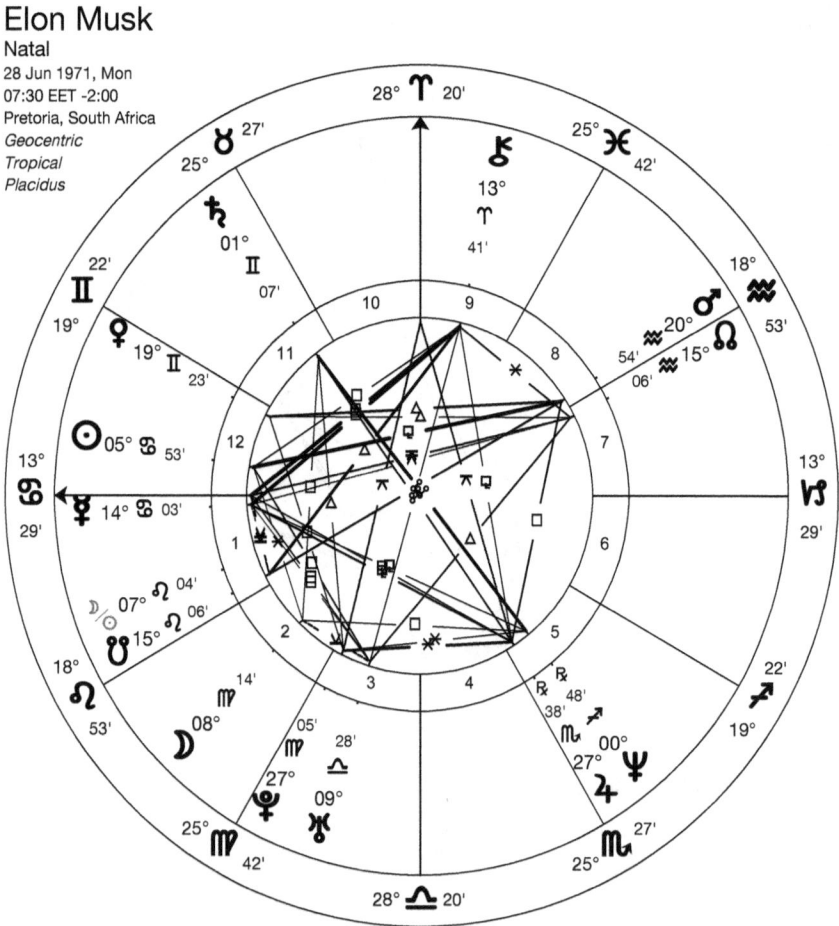

Figure 8.14.9

with Mars at 27 Scorpio on his natal Jupiter, later aligning with Mercury at 13 Capricorn on his Descendant. (Figure 8.14.9)

Space

Musk's aerospace business SpaceX also made a breakthrough during this period, sending its first civilians to the International Space Station. Shortly afterwards Amazon founder Jeff Bezos launched his own space tourism operation Blue Horizon. Russia sent up a film production crew for the first time.

Other breakthroughs were:

- NASA flying a drone on Mars, the first powered flight on another planet;
- China starting to build its first space station;
- NASA successfully testing its ability to alter the orbit of an asteroid;
- NASA recommencing missions to the Moon using the uncrewed Artemis 1 spacecraft;
- The Event Horizon Telescope revealing the first image of a black hole at the centre of our galaxy, and confirming Einstein's theory of relativity.

On Christmas Day 2021 while Comet Leonard was passing through Aquarius, NASA launched the largest, most powerful and complex telescope ever built to help astronomers gain a better understanding of the origins of the universe. The first image taken by the James Webb Space Telescope was revealed the following July, capturing thousands of galaxies some of which were 13 billion years old, the oldest ever seen.

It is interesting that the arrival of a comet from so far away that took tens of thousands of years to get here should coincide with the launch of a satellite that allows us to see further away in space and further back in time than we have ever been able to do before.

Cryptocurrencies

2021 was a highly volatile year for Bitcoin, largely thanks to the behaviour of our old friend Elon Musk. At the start of the year he added #bitcoin to his Twitter handle, pushing its value up from $5,000 to $37,500 in a single hour.

Confidence in crypto also rose during that summer when the Swiss canton Zug allowed people to pay tax using cryptocurrencies, and El Salvador became the first country to accept Bitcoin as legal tender. However in 2022 Bitcoin's value plummeted from $69,000 to just $16,000 after governments in China and the US started to introduce state controls. Crypto exchanges started to go down, banning withdrawals and laying off staff with several going bankrupt.

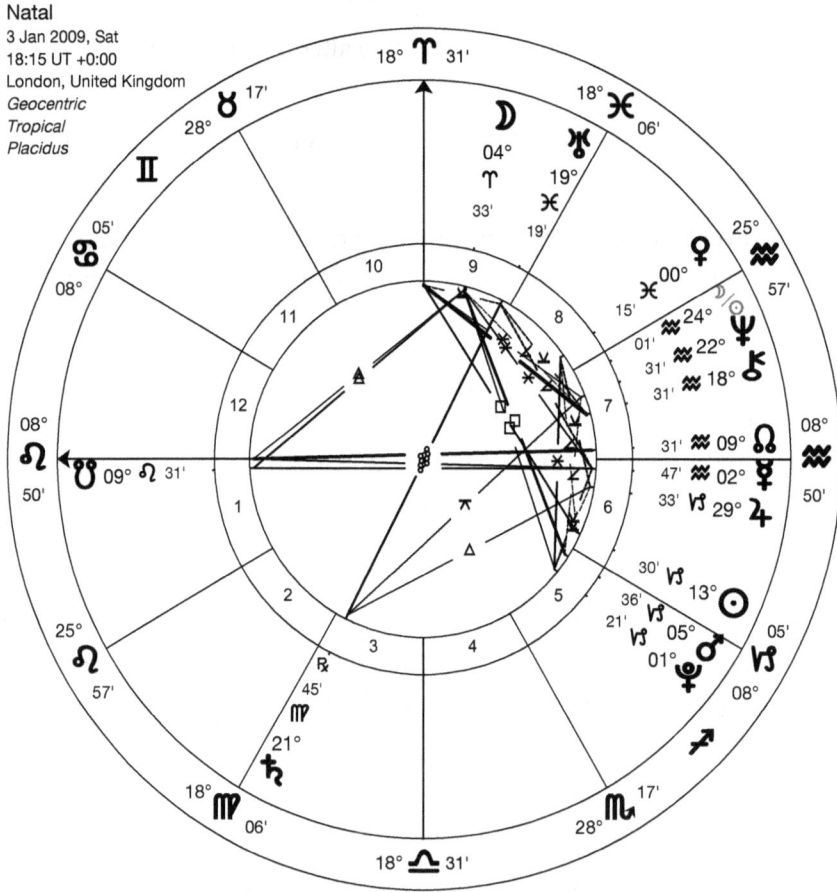

Figure 8.14.10

The world's third largest crypto exchange FTX valued at $32 billion declared bankruptcy. Along with colleagues, founder Sam Bankman-Fried was arrested and charged on seven counts of fraud and money laundering. He was given a 25 year prison sentence and ordered to forfeit $11 billion.

Comet Leonard reached perihelion at 15 Aquarius on Bitcoin's Chiron and not far from its Neptune. It aligned with Mercury at 13 Capricorn, the degree of Bitcoin's Sun and just a few degrees from where it crossed the ecliptic and aligned with fixed star Albadah. It also aligned with transiting Venus at 6 Capricorn, one degree from Bitcoin's Mars at 5 Capricorn, and with transiting Pluto at 27 Capricorn close to Bitcoin's Jupiter. (Figure 8.14.10)

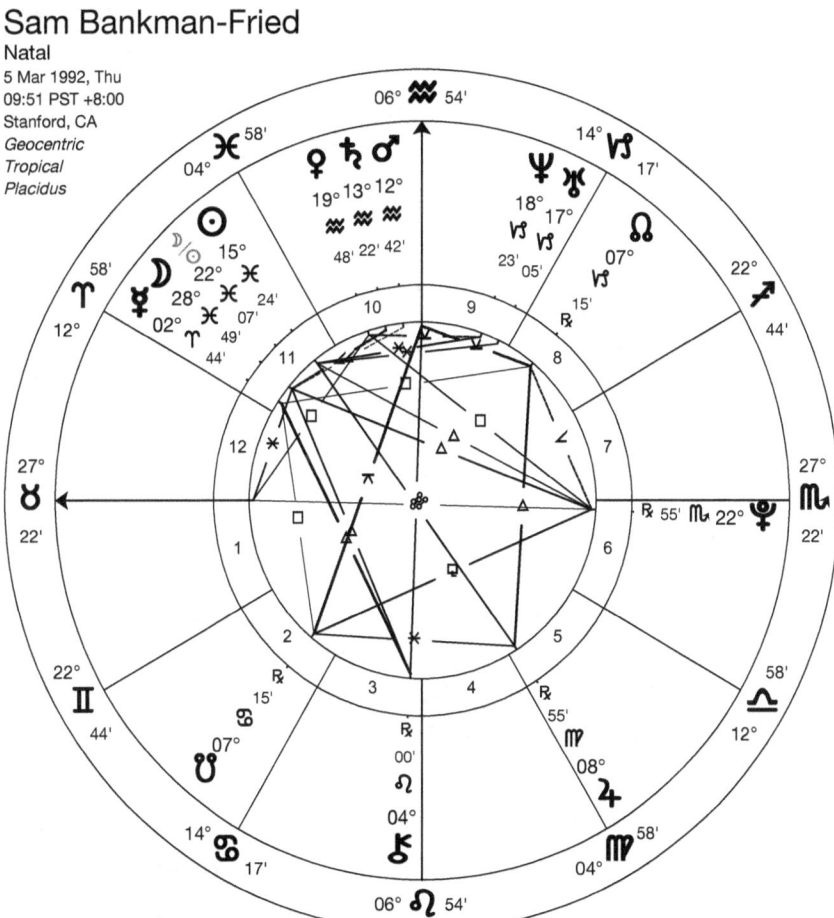

Figure 8.14.11

166   *Comets in Astrology*

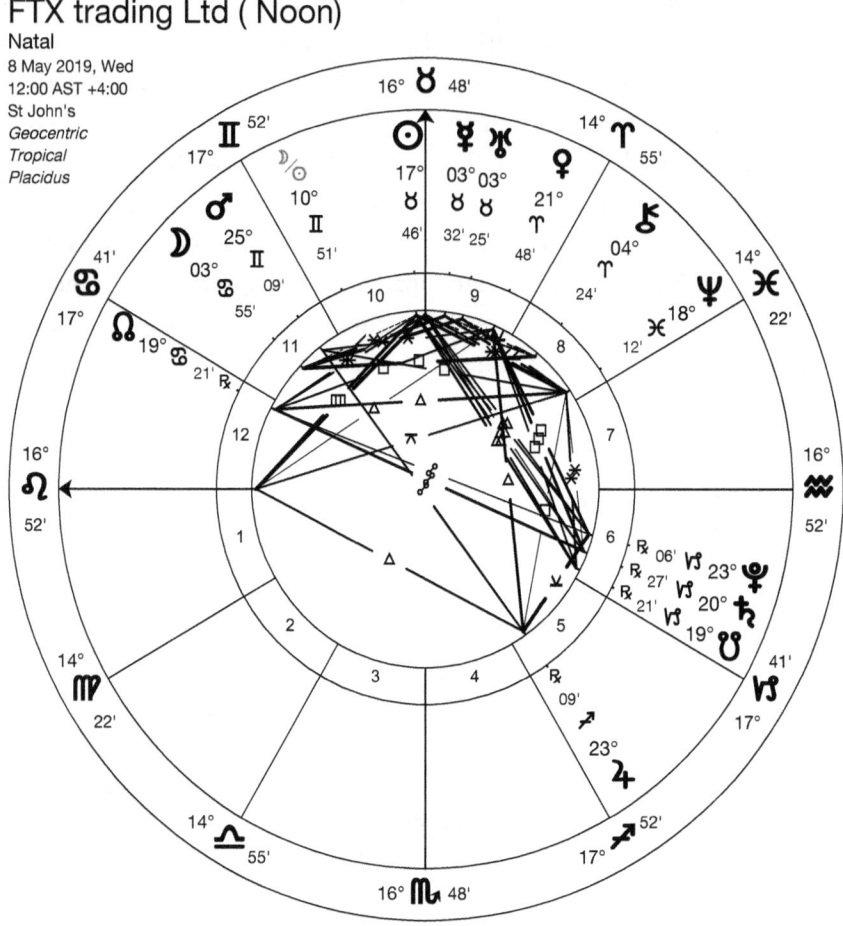

Figure 8.14.12

The comet crossed the ecliptic at 17 Capricorn right on Sam Bankman-Fried's natal Uranus-Neptune conjunction, which is in turn very close to the South Node in the chart for the founding of his company FTX at 19 Capricorn. The comet reached perihelion and disappeared from view at 15 Aquarius, just after crossing Bankman-Fried's natal Mars-Saturn conjunction. (Figures 8.14.11 and 8.14.12)

Comet Leonard through the elements

Comet Leonard travelled from 18 Libra to 15 Aquarius, moving through all four astrological elements on its path, all of which were activated in an extreme way during this period

Air

Climate continued to rise up the political agenda with the US rejoining the Paris Climate Agreement as the International Panel on Climate Change warned the planet was racing past the 1.5C threshold considered critical to life on planet Earth.

Those warnings came as extreme weather events continued to plague the planet. In North America a four day winter storm killed 136 people and left 9.9 million without power in 2021, and the following year another winter storm killed 91 in the US and Canada. Hundreds died in some of the worst cyclones, typhoons, hurricanes and tropical storms ever experienced in North and South America, east Asia and southern Africa.

Water

Climate change was also blamed for causing the extreme flooding during this period. Around 230 people lost their lives in floods in Germany and Belgium, 11 died in Madagascar, but worst hit was Pakistan where over 1,000 people died due to flooding in August 2022.

In March 2021 the massive cargo ship Ever Given blocked the Suez Canal, one of the world's most critical trade routes, after being blown sideways by high winds and a dust storm. The blockage lasted six days leaving more than 350 ships queuing up to go through the canal, that carries around $10 billion worth of goods every day.

A month later the Japanese government authorised the slow release of treated radioactive water from the disabled Fukushima nuclear plant, causing alarm in environmental circles. That same month an Indonesian navy submarine sank during an exercise with the loss of all 53 crew.

Fire

Record temperatures sparked wildfires throughout Europe and the Mediterranean affecting 25 different countries. Around 53,000 people lost their lives and tens of thousands had to evacuate their homes. Even the temperate UK had nearly 25,000 wildfires that year, while in Canada 60 people died after lightning strikes coupled with the heatwave sparked 130 separate wildfires.

The presidents of Japan, Haiti and Chad were assassinated as was the Italian ambassador to the Democratic Republic of Congo and controversial author Salman Rushdie almost died when he was stabbed at a public lecture in New York.

There were military coups in Armenia, Mali, Guinea, Sudan, Burkina Faso and Myanmar, where democracy leader Aung San Suu Kyi was arrested shortly after winning an election, and in the US Trump supporters stormed the Capitol building.

Israel continued its assault on the Palestinian population of Gaza and attacked Iran, blowing up a uranium enrichment plant. Deadly clashes broke out on the borders of central Asian countries Kyrgyzstan and Tajikistan, and also between Armenia and Azerbaijan. There was fighting in the Congo and Ethiopia, and the terror group Islamic State left hundreds dead in Pakistan, Nigeria and Somalia.

The Taliban took back control of Afghanistan and the US pulled out in a hurry after occupying the country for almost 20 years, leaving behind huge amounts of military equipment and an uncertain future for the Afghan people.

Earth

The most powerful volcano since the Krakatoa eruption of 1883 took place in Tonga, causing a tsunami with waves up to 45 metres high. Scientists said it was the largest atmospheric explosion ever recorded on modern instruments, hundreds of times more powerful than the Hiroshima atomic bomb, and was felt across the entire coastline of the Pacific Ocean from New Zealand to Alaska, Peru to Russia.

More deadly were the earthquakes in Haiti that killed 2,500 people, on the border between Afghanistan and Pakistan where 1,163 died, in China with 117 deaths and in Indonesia where 635 lost their lives.

An earthquake in the world of finance took place when the International Consortium of Investigative Journalists released the Pandora Papers exposing offshore accounts with a combined value of trillions of dollars belonging to 35 world leaders, including current and former presidents, prime ministers and heads of state, as well as more than 100 business leaders, billionaires and celebrities. Figures named include former UK prime minister Tony Blair, Ukraine president Volodymyr Zelensky, King Abdullah the II of Jordan, supermodel Claudia Schiffer and singers Shakira and Julio Iglesias.

Comet Case Studies 169

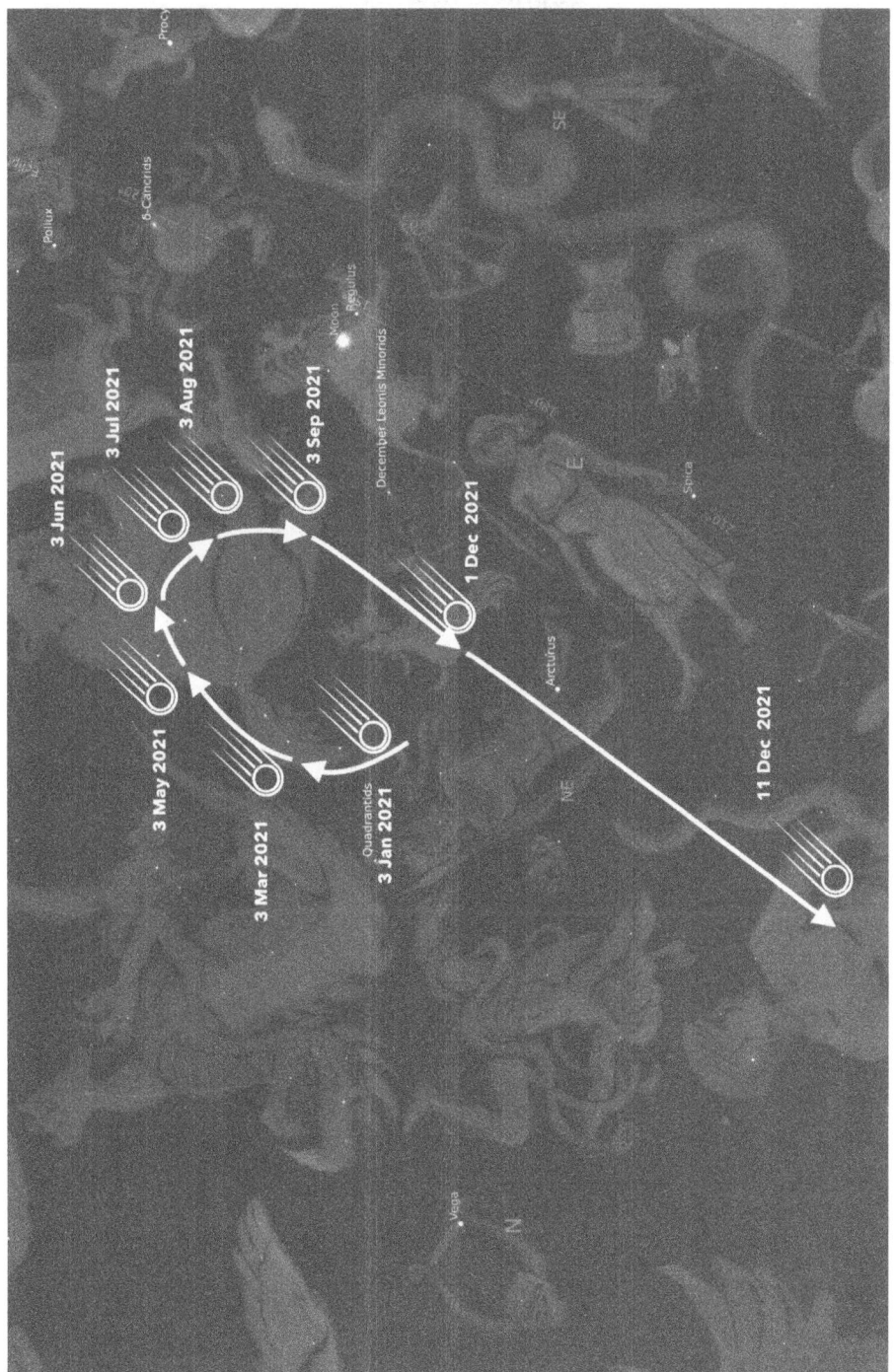

The trajectory of Comet Leonard in 2021

Comets of 2024
Comet 12P Pons-Brooks

Type of comet: periodic

Orbital period : 71 years

Astronomical data

	Date	Position
Discovery	12 Jul 1812	23 Gemini
First seen	20 Mar 2024	28 Aries
Last Perihelion	21 Apr 2024	22 Taurus
Last seen	25 May 2024	18 Gemini
Crossed celestial equator	04 May 2021 22:46 UT	02 Gemini
Crossed ecliptic	11 Apr 2024 10:54 UT	15 Taurus
Appearance during solar eclipse	8 Apr 2024	13 Taurus

Tropical zodiac

Dates	Sign
20 Mar 2024 - 22 Mar 2024	Aries
22 Mar 2024 - 02 May 2024	Taurus
02 May 2024 - 25 May 2024	Gemini

Constellations

Note: not to be confused with the 12 signs of the tropical zodiac.

Date	Constellation
20 Mar 2024 - 27 Mar 2024	Pisces
27 Mar 2024 - 19 Apr 2024	Aries
19 Apr 2024 - 04 May 2024	Taurus
04 May 2024 - 20 May 2024	Eridanus
20 May 2024 - 25 May 2024	Lepus

Fixed stars

Date	Star
31 Mar 2024	Hamel

Planetary alignments

Date	Planet	Location
19 Apr 2024	Jupiter	21 Taurus
20 Apr 2024	Uranus	21 Taurus

Comet C/2023 A3 Tsuchinshan-ATLAS

Comet type: long periodic/hyperbolic

Orbital period: millions of years

Astronomical data

	Date	Position
Discovery	09 Jan 2023	08 Scorpio
First seen	12 Sep 2024	13 Virgo
Perihelion	27 Sep 2024	15 Virgo
Last seen	02 Nov 2024	01 Capricorn
Crossed celestial equator	14 Oct 2024 11:11 UT	09 Scorpio
Crossed ecliptic	08 Oct 2024 04:59 UT	09 Libra

Tropical Zodiac

Dates	Sign
27 Aug 2024 - 05 Oct 2024	Virgo
05 Oct 2024 - 12 Oct 2024	Libra
12 Oct 2024 - 18 Oct 2024	Scorpio
18 Oct 2024 - 30 Oct 2024	Sagittarius
30 Oct 2024 - 10 Nov 2024	Capricorn

Constellations

Note: not to be confused with the 12 signs of the tropical zodiac.

Date	Constellation
27 Aug 2024 - 28 Sep 2024	Sextans
28 Sep 2024 - 04 Oct 2024	Leo
04 Oct 2025 - 15 Oct 2024	Virgo
15 Oct 2024 - 19 Oct 2024	Serpens
19 Oct 2024 - 10 Nov 2024	Ophiuchus

Planetary alignments

Date	Planet	Zodiac location
17 Sep 2024	Mercury	12 Virgo
09 Oct 2024	Sun	17 Libra
11 Oct 2024	Mercury	25 Libra
19 Oct 2024	Venus	00 Sagittarius

Events

The year 2024 saw two significant comets pass through the heavens, both with very different characteristics. The first one to appear was Comet 12P/Pons-Brooks, a periodic comet with an orbital period of 71 years, that was discovered in 1812 by French astronomer Jean-Louis Pons and again in 1883 by British-born American astronomer William Robert Brooks. However historical records suggest it was recorded as far back as 1313 in east Asia. The last time it was seen was in 1953/4, from when certain themes can be observed. Events from then include:

- The death of Josef Stalin, the USSR withdrawing from Austria and handing Crimea to Ukraine;
- The discovery of DNA;
- The first mass vaccination roll out to combat polio;
- The first rock 'n' roll records 'Rock Around the Clock' and Elvis Presley's 'That's Alright';
- The overthrow of Iran's first democratically elected leader;
- The end of the Korean War and the start of fighting in Vietnam;
- The first nuclear submarine, first Soviet nuclear bomb and the first Pacific nuclear tests;
- The first transistor radio;
- Elizabeth II was crowned.

In contrast, Comet C/2023 A3 (Tsuchinshan-ATLAS) will not return for millions of years, if ever, yet its brightness raised considerable excitement in the astronomical community, leading some to call it "the comet of the century". It is now known as the Great Comet of 2024.

As the two comets appeared in the spring and autumn of 2024, we can assume their impact would also have been felt during 2023 and 2025.

Gaza

With the war in Ukraine entering its second year, the theme of conflict exploded even more with the events of 7 October 2023 when Hamas staged their attack on Israel's southern border resulting in over 1,000 deaths and the capture of 251 hostages. Israel's response saw tens of thousands of Palestinians killed, the levelling of large parts of the Gaza Strip and charges of genocide. This in turn led to the conflict expanding to take in Hezbollah

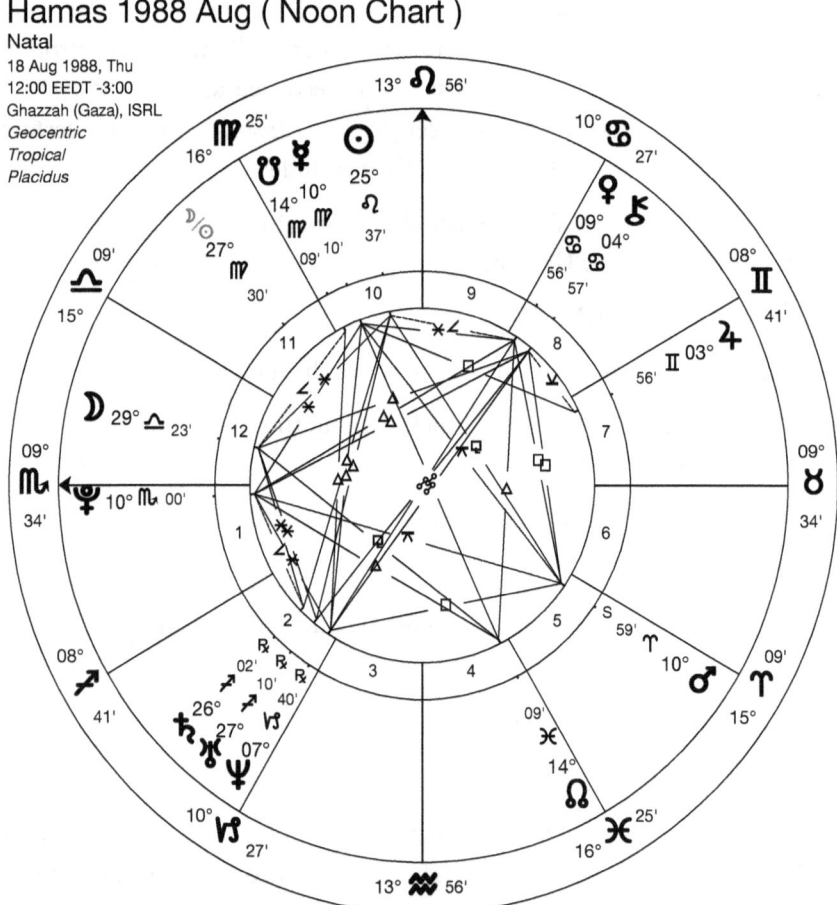

Figure 8.15.1

in Lebanon, the Houthis in Yemen and even Iran who all targeted Israel with missiles.

Just two months before the 7 October attack, long period Comet Nishimura (C/2023 P1) was discovered and could be seen by the naked eye by September. It appeared suddenly before sunrise in the fire sign of Leo, a clear forewarning of war.

We can see from the chart for the Covenant of the Islamic Resistance Movement, known as Hamas, on 18 August 1988, that it has the Sun at 25 Leo, one degree from where the comet was first seen at 24 Leo. (Figure 8.15.1)

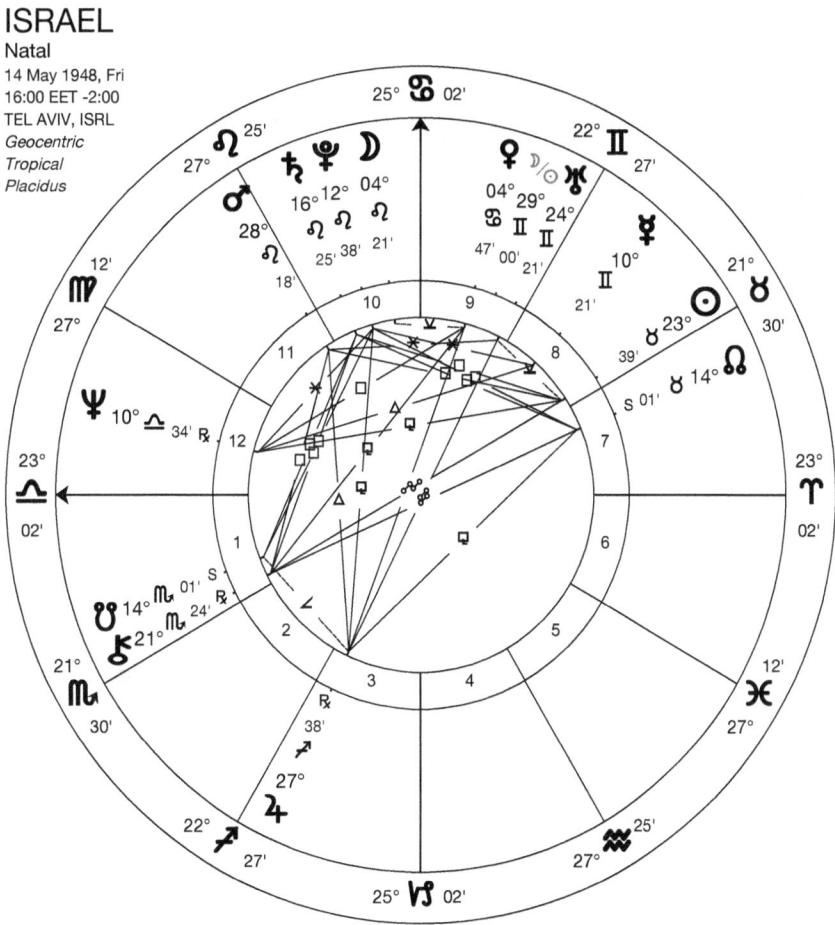

Figure 8.15.2

Understanding the Gaza conflict astrologically leads one to focus on the ingress and eclipse charts for 2023, but in this book we are concentrating on the influence of comets.

On 31 March 2024 Comet 12P Pons-Brooks aligned with the fixed star Hamal – strangely similar to the name 'Hamas' – also known as Alpha Arietis the forehead of the Ram in the Aries constellation. According to the ancient astrologer Ptolemy this star has the nature of Mars and Saturn and symbolises violence and aggression. In its positive aspect it can symbolise protecting the family and community.

The triple conjunction of Comet Pons-Brooks with Jupiter and Uranus at 21 Taurus was just 2 degrees from the Sun in the chart for Israel. It was discovered at 23 Gemini very close to Israel's Uranus and crossed the celestial equator at 2 Gemini very close to Jupiter in the Hamas chart. (Figure 8.15.2)

Meanwhile Comet Tsuchinshan-ATLAS was discovered at 8 Scorpio and crossed the equator at 9 Scorpio moving rapidly towards an exact conjunction with Pluto in the Hamas chart at 10 Scorpio.

Earlier on 17 September it had aligned with transiting Mercury at 12 Virgo, very close to Mercury and the South Node in the Hamas chart. It aligned with transiting Mercury a second time at 25 Libra on 11 October right on Israel's Ascendant having crossed the ecliptic at 9 Libra very close to Israel's Neptune.

This double alignment with Mercury in both charts may emphasise the importance of communication and messaging for both sides of this conflict. It is the first of its kind to be transmitted live to people's mobile phones all over the world.

US election

2024 saw 64 elections take place around the world, indicating what a year of potential change it offered the political realm. While some countries, notably Russia and China, saw no change resulting from their elections, the US witnessed a complete transformation in leadership style from Democrat Joe Biden to Republican Donald Trump. The election campaign itself saw high drama with more than one assassination attempt on Trump's life, and Biden stepping down from the leadership race handing the baton over to his deputy Kamala Harris.

Looking at the Aries ingress chart for the year, Mars was on the IC at 27 Aquarius squaring Jupiter and Uranus at 21 Taurus on the Descendant. The Sibly chart for the formation of the US in 1776 has the Moon at 27 Aquarius, suggesting the people may be agitated and feel under attack. This would have been triggered in July, the month Trump narrowly dodged a bullet, when Mars and Uranus conjoined at 26 Taurus squaring Mars in the ingress chart and the Moon in the Sibly chart. (Figure 8.13.3)

Regarding our two 2024 comets, Pons-Brooks was discovered at 23 Gemini, very close to Trump's natal Sun-North Node conjunction, and was last seen at 18 Gemini very close to his natal Uranus. On 20 April the comet aligned with the transiting Jupiter-Uranus conjunction at 21 Taurus close to his MC. (Figure 8.13.1)

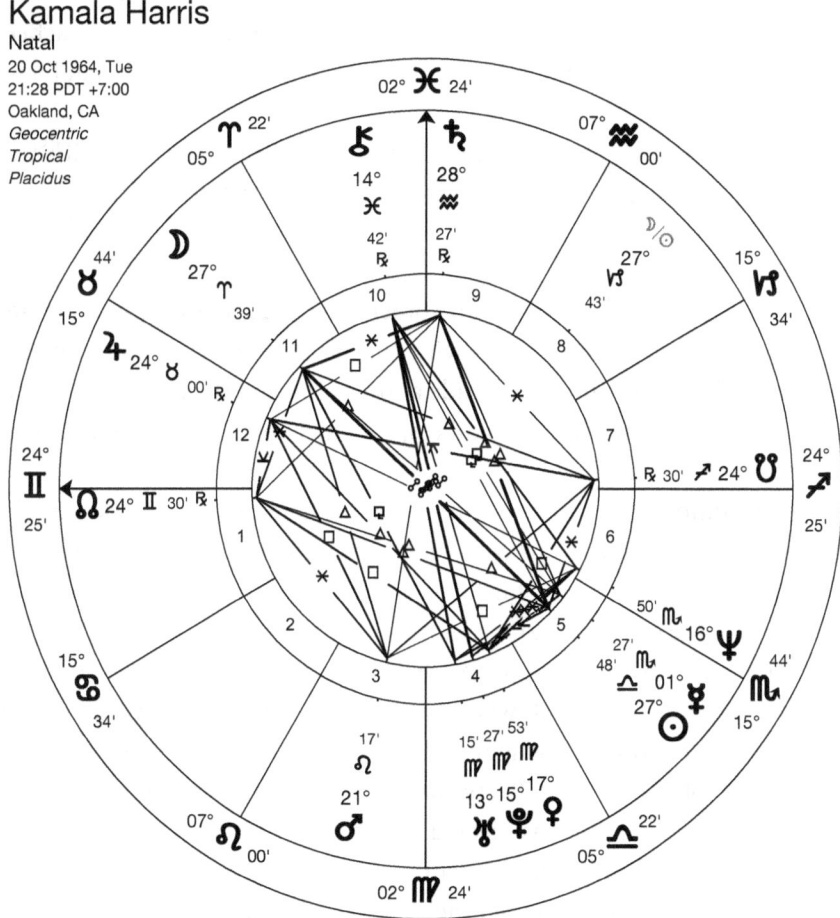

Figure 8.15.3

For Biden, the comet reached perihelion at 22 Taurus, directly opposite his natal MC ruler Mercury at 22 Scorpio. It crossed the celestial equator at 2 Gemini on his natal Uranus and Descendant, and was first seen at 28 Aries, close to his 8th house-ruling Moon at 0 Taurus, all of which can be interpreted to reflect his change of status during the campaign. (Figure 8.13.2)

The comet's impact on Harris looked more positive. Its 23 Gemini discovery point was on her natal Ascendant, it was first seen by naked eye at 28 Aries on her natal Moon and aligned with the transiting Jupiter-Uranus conjunction just three degrees from her natal Jupiter. The good fortune may have been enough to make her the Democrat candidate, but it was not enough to win her the election. (Figure 8.15.3)

Comet Tsuchinshan-ATLAS aligned with her natal Uranus when it was first seen at 13 Virgo, with natal Pluto and Venus when it reached perihelion at 15 Virgo and with transiting Mercury close to her natal Sun at 25 Libra.

Its impact on Trump was more straightforward. It aligned with the transiting Sun at 17 Libra on 9 October, exactly on Trump's Jupiter.

Canada

The period impacted by these two comets saw sudden, unexpected shifts in Canada's political leadership with the resignation of prime minister Justin Trudeau after almost a decade at the helm, and the election of Mark Carney, one of the world's top bankers, who had never been an elected MP before.

At a mundane level, these shifts were triggered by the election of US president Donald Trump in November 2024, followed by his announcement that he would impose tariffs on trade with Canada and his desire to incorporate the country into the USA as its 51st state.

Trudeau's reaction was to invite Carney to take over from finance minister Chrystia Freeland, who responded by resigning very publicly in December 2024. The following month Trudeau resigned as prime minister and leader of the Liberal Party amid plummeting polls and regional election losses. Carney was elected party leader in March and became prime minister of a minority government in April 2025.

Comet Pons-Brooks crossed the celestial equator at 2 Gemini, close to Trudeau's Saturn at 0 Gemini, while Comet Tsuchinshan-ATLAS aligned with the Sun at 17 Libra right on his natal Uranus, and with Venus at 0 Sagittarius, close to his natal Neptune – all indicating shock and destabilisation. Perhaps most significant though was the comet disappearing at 1 Capricorn in early November 2024 – the time of Trump's election – just 2 degrees from his natal Sun. (Figure 8.15.5)

Carney's election success was a surprise, as the Liberal Party had been trailing badly in the polls. However, having led the Bank of Canada through the 2008 global financial crisis and the Bank of England through the Covid-19 pandemic, he was seen as a safe pair of hands as Canada faced an apparent economic and political assault from the US. The planets were on Carney's side at the time of his election with Uranus at 25 Taurus

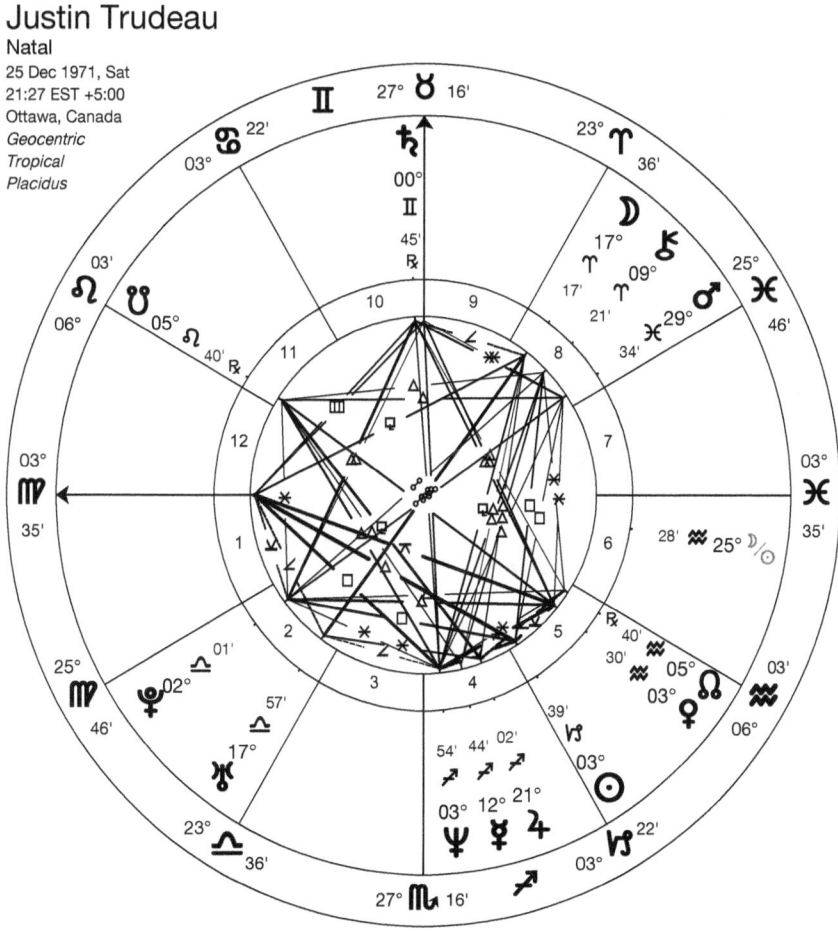

Figure 8.15.5

conjunct his natal Ascendant and Jupiter at 20 Gemini at his Sun-Moon midpoint. (Figure 8.15.6)

But both comets had a huge influence on Carney's natal chart. Comet Pons-Brooks was discovered at 23 Gemini near his Sun-Moon midpoint and was last seen at 17 Gemini, one degree from his North Node. Perhaps more significantly, the comet aligned with the Jupiter-Uranus conjunction in April 2024 at 21 Taurus, just one degree from his natal Jupiter at 22 Taurus and close to his 25 Taurus Ascendant – suggesting sudden changes could bring success and expand his influence.

Not only that, Comet Tsuchinshan-ATLAS was first seen at 13 Virgo and reached perihelion at 15 Virgo, having aligned with transiting Mercury

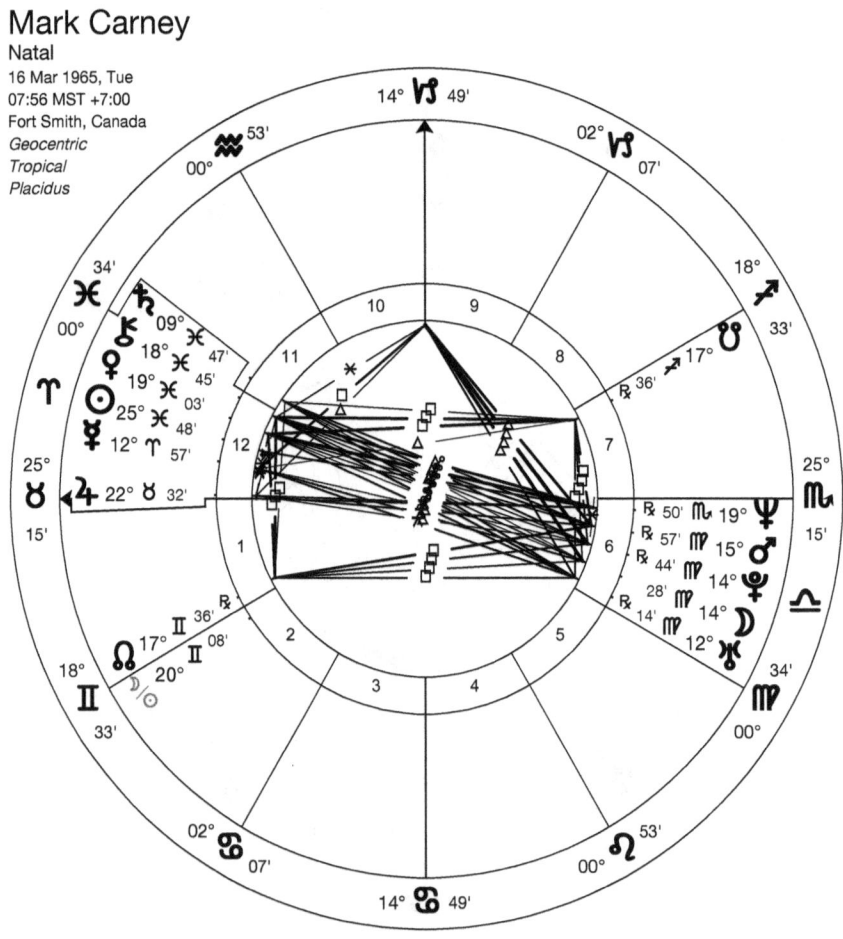

Figure 8.15.6

Comet Case Studies 181

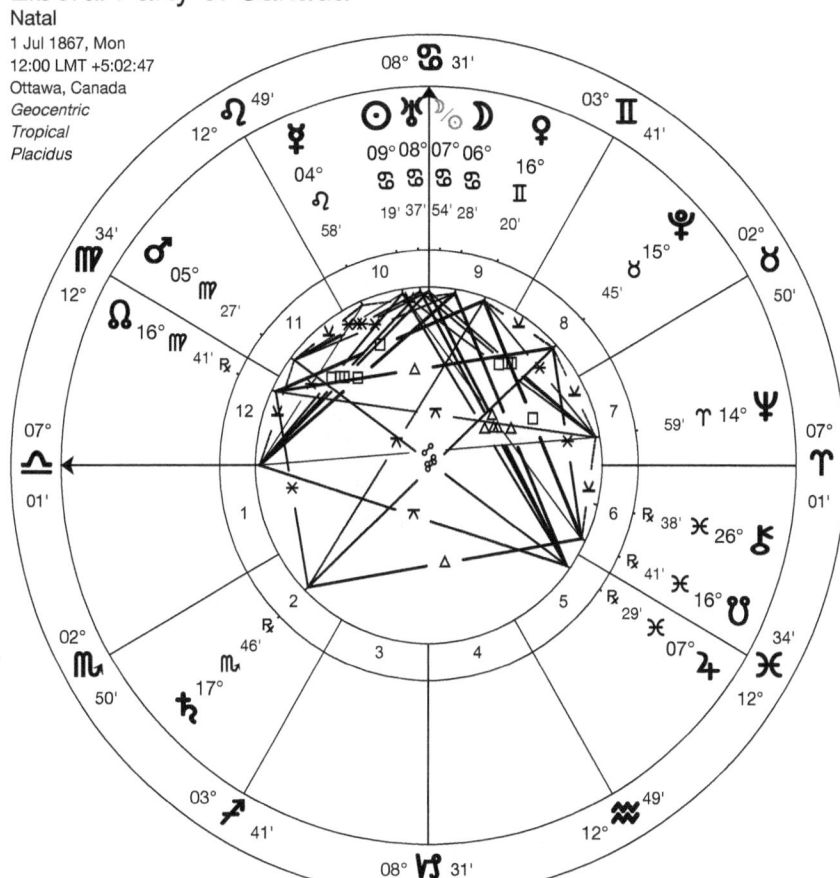

Figure 8.15.7

at 12 Virgo. Imagine the impact on Carney's natal four planet stellium – Uranus, Moon, Pluto and Mars between 12 and 15 Virgo.

Finally we can see both comets impacting the chart of Canada's Liberal Party, founded on 1 July 1867. Comet Pons-Brooks crossed the ecliptic at 15 Taurus, right on Pluto in the chart. It was last seen at 18 Libra, where Venus resides in the chart, symbolising the party's popularity.

Comet Tsuchinshan-ATLAS reached perihelion at 15 Virgo conjunct the North Node in the party's chart. In mundane astrology, the nodal axis symbolises connection with the people. (Figure 8.15.7)

Comets through the elements

Between them, the two comets passed through all four elements.

Fire

The period from 2023-2025 has seen widespread military conflict and civil unrest. Aside from the escalation and intensity of the Ukraine war and the devastation of Gaza in Palestine, which drew in Lebanon, Iran and the Houthis in Yemen, there was a series of coups and conflicts raging in sub-Saharan Africa, including the Democratic Republic of Congo, Sudan, Chad, and Gabon. The period also saw the French military being forced to leave countries like Nigeria and Burkina Faso after leadership changes.

A deadly civil war continued in Myanmar, the conflict between Armenia and Azerbaijan over Nagorno-Karabakh flared up again, ethnic violence troubled India, a series of mass shootings, stabbings and terror attacks took place in the US, Canada, western and eastern Europe and a war over Kashmir between India and Pakistan became an increasing threat.

As well as the assassination attempt on Trump, the Slovakian prime minister Robert Fico survived being shot, while the CEO of UnitedHealthcare Brian Thompson did not survive being shot, and Iranian president Ebrahim Raisi died in a helicopter crash.

The period saw the highest post-industrial temperatures ever recorded with massive wildfires striking Canada, Greece, Hawaii, Los Angeles and South Korea with many others taking place all over the world.

In Macedonia a nightclub fire killed 569 people, 77 died in a house fire in South Africa and a fire burned for two days in Ohio after a train carrying toxic chemicals crashed.

Earth

The death toll from earthquakes during this period was extremely high, with some of the most deadly of this century so far occurring in Morocco with the loss of almost 3,000 lives and Myanmar where almost 5,500 died. Earthquakes killed 1,400 in Afghanistan, over 500 in Japan, and over 1,200 in Tibet, while landslides killed around 2,000 in Papua New Guinea, over 250 in Ethiopia and more than 330 in India.

The biggest banking crisis since 2008 saw two major US banks collapse, record interest rate rises in Europe to tackle runaway inflation, multi-billion dollar takeovers in the oil, tech and fashion industries and a growing trend away from the dollar to settle international trade deals in the face of growing sanctions by the US.

The move away from fossil fuels was agreed by the world's leading nations in the G7, while the UK and Europe agreed to phase out coal power by 2035.

Air

Records fell left, right and centre during these years around wind speeds, deaths and damage with a huge number of extreme weather events. Cyclone Freddy was the longest lasting cyclone in recorded history to affect the Indian Ocean, killing 1,400 people in Malawi and Mozambique. Storm Daniel was the worst cyclone to ever hit the Mediterranean causing flooding in Greece, Bulgaria, Turkey and Libya where two dams collapsed resulting in 5,900 deaths, 7,000 injuries and leaving up to tens of thousands of people missing.

The third largest outbreak of tornadoes caused $5.4 billion of damage in the US, south east Asia experienced its costliest ever typhoon, Mexico was hit by its most powerful cyclone, the island of Granada had its earliest ever hurricane and southern California had record flooding from Hurricane Hilary. Hundreds more died and billions of dollars more in damage was recorded in America, east Africa and south Asia. And Ireland experienced 114mph winds with Storm Eowyn, cutting power for one million homes.

The race into space continued apace with India landing on the Moon and launching its first mission to observe the Sun, Japan landing on the Moon, the first commercial company landing on the Moon, and the highest altitude spacewalk since the US Apollo programme in the '70s.

Two missions to explore the moons of Jupiter were launched by NASA and the European Space Agency, while the period saw significant growth in private sector space travel with SpaceX launching its biggest ever rocket, the Starship, Boeing launching Starline, and Amazon boss Jeff Bezos launching his venture Blue Horizon.

We also saw the first close up image of a star outside the Milky Way, the closest ever flight to the Sun, a form of life discovered on a distant planet for the first time and the first ever combat in space when Israel shot down a missile launched from Yemen by the Houthis.

Water

Floods were a major feature of this period causing hundreds of deaths and making thousands homeless in Congo, the Persian Gulf, Brazil, Kenya, Tanzania and central Europe, while eastern Spain saw a year's worth of rain fall in just eight hours killing more than 230 people.

Migrants continued to risk their lives escaping to Europe on overcrowded boats, in one case 500 were left missing and 82 were drowned when their boat capsized off Greece. There was a high profile incident when five people on board a submersible imploded on a tourist trip to view the wreck of the Titanic.

Meanwhile environmental campaign group Greenpeace declared "the biggest conservation victory ever" with the United Nations signing the High Seas Treaty to protect marine biological diversity in waters outside national jurisdiction through which they hope to protect 30 per cent of the world's oceans.

Conclusion

This book is a compilation of my research into the astrological significance of comets carried out over the past twelve years. I hope I have succeeded in demonstrating the impact comets can have on both our personal lives and on the mundane world, especially when they cross sensitive points in the sky while visible to the naked eye.

When I started out on this journey I was already aware that throughout history comets have been regarded as bad omens, and I have been surprised to discover they can also be associated with release and major breakthroughs. I like to think of the example in Greek mythology of Electra's grief over the fall of Troy and the dream warning Aeneas to flee the falling city as a comet appeared in the sky and then going on to found a new Rome. We too can view comets as signs to leave the past behind and create a new future.

In China comets are called "hui xin". The word "hui" is equivalent to "wisdom" in English, and "xin" means heart or mind. In Chinese we say: "Wield the sword of wisdom to cut away attachments," and we can easily imagine comets symbolising swords in the sky, not least by the shape they form with their bright tails. Could we look upon the appearance of a comet as symbolising the birth of a new wisdom? Could their position indicate where to seek such wisdom in our lives? Could they offer us an opportunity to cut through bonds and attachments to allow fresh new experiences into our lives?

Since I began my research I have been very happy to see many of my astrological colleagues join me in considering the meaning of comets. We are just at the beginning of this journey in modern astrology and my hope is that the astrological community will devote more time and energy into researching this fascinating field of study.

Comets will keep appearing, I will continue paying attention to them and will share my observations on my website at www.rodchang.com

Thank you for reading and I hope you will join me on this exciting journey of discovery.

Notes & References

Chapter 1
[1] Oxford English Dictionary
[2] https://www.etymonline.com/word/comet
[3] Shuowen Jiezi Vol 4 Dingyuan Culture Publishing 2008
[4] https://skyandtelescope.org/astronomy-resources/what-is-a-comet/
[5] https://en.wikipedia.org/wiki/2060_Chiron
[6] https://skyandtelescope.org/astronomy-resources/what-is-a-comet/
[7] Hughes, D. W. Journal of the British Astronomical Association, vol.101, no.2, p.119-120
[8] https://en.wikipedia.org/wiki/Active_asteroid
[9] https://en.wikipedia.org/wiki/Comet
[10] Greenberg, J. Mayo Astronomy and Astrophysics, v.330, p.375-380 (1998)
[11] DNA building blocks can be made in space, by NASA's Goddard Space Flight Center, Greenbelt, Maryland https://www.astronomy.com/science/dna-building-blocks-can-be-made-in-space/
[12] https://en.wikipedia.org/wiki/Comet
[13] https://en.wikipedia.org/wiki/Great_comet
[14] https://en.wikipedia.org/wiki/12P/Pons%E2%80%93Brooks
[15] https://en.wikipedia.org/wiki/Comet_Hale%E2%80%93Bopp
[16] https://en.wikipedia.org/wiki/Comet
[17] https://en.wikipedia.org/wiki/Meteor_shower
[18] https://en.wikipedia.org/wiki/Comet_Kohoutek

Chapter 2
[19] Autumn Annals https://ctext.org/chun-qiu-zuo-zhuan/wen-gong/zh
[20] Sima Qian, The Records of the Grand Historian https://ctext.org/shiji/qin-shi-huang-ben-ji/zh
[21] Donald Yeomans, Comets: A Chronological History of Observation, Science, Myth and Folklore. Wiley Science Editions, 1991.
[22] Aristotle, Meteorologica Book 1 Chapter 7.
[23] Manilius, Astronomica. Edit. and translated by G.P Goold 1977
[24] https://en.wikipedia.org/wiki/Julius_Caesar

25. Pliny the Elder, The Natural History. Translated by John Bostock, https://www.perseus.tufts.edu/hopper/text?doc=Perseus%3Atext%3A1999.02.0137%3Abook%3D2%3Achapter%3D22
26. https://en.wikipedia.org/wiki/Seneca_the_Younger
27. Pliny the Elder, The Natural History. Translated by John Bostock, https://www.perseus.tufts.edu/hopper/text?doc=Perseus%3Atext%3A1999.02.0137%3Abook%3D2%3Achapter%3D22
28. https://en.wikipedia.org/wiki/Halley%27s_Comet
29. https://en.wikipedia.org/wiki/Great_Comet_of_1577
30. Diarium astrologicum et metheorologicum anno a nato Christo 1586. Et de cometa quodam rotundo omnique cauda destituto, qui anno proxime elapso, mensibus Octobri & Novembri conspiciebatur, ex observationibus certis desumta consideratio astrologica. Uraniborg: Officina Uraniburgica, 1586.
31. https://en.wikipedia.org/wiki/Great_Comet_of_1680
32. https://en.wikipedia.org/wiki/Halley%27s_Comet
33. Edwin Emerson, Halley's Comet in History and Astronomy, 2017

Chapter 3

34. Tycho Brahe, Diarium astrologicum et metheorologicum anni a nato Christo 1586 Uranienborg: 1586
35. William Lilly, New Prophecy, Or, Strange and Wonderful Predictions relating to the year, 1678. As well as from the Great Blazing-Star [London] 1678, 3-4]
36. Ptolemy, Tetrabiblos Ch 10 'Colours in Eclipses, Comets, and Similar Phenomena'
37. John Gadbury, De Cometis: or, a Discourse of the Natures and Effects of Comets
38. John Bainbridge, An Astronomical Description of the Late Comet, p.36, 1619
39. Tycho Brahe, Diarium astrologicum et metheorologicum anni a nato Christo 1586 Uranienborg: 1586.
40. Tychonis Brache, Dani opera ominia Copenhagen 1922
41. William Lilly, New Prophecy, Or, Strange and Wonderful Predictions relating to the year, 1678. As well as from the Great Blazing-Star [London] 1678, 3-4]
42. Tycho Brahe Diarium astrologicum et metheorologicum anni a nato Christo 1586 Uranienborg: 1586
43. Ptolemy The Centiloquy 235
44. William Lilly, Christian Astrology Book 1 Ch 8 -14 1647
45. Sara Schechner Genuth, Comets, Popular Culture & the Birth of Modern Cosmology, 1997
46. https://www.forbes.com/sites/startswithabang/2018/08/09/this-is-why-comets-glow-an-eerie-green-color/
47. https://cloudbreakoptics.com/blogs/news/astrophotography-pixel-by-pixel-part-5-one-shot-color-and-monochrome-sensors
48. https://cloudbreakoptics.com/blogs/news/astrophotography-pixel-by-pixel-part-5-one-shot-color-and-monochrome-sensors

188 Comets in Astrology

49 Pliny the Elder, The Natural History. Translated by John Bostock, https://www.perseus.tufts.edu/hopper/text?doc=Perseus%3Atext%3A1999.02.0137%3Abook%3D2%3Achapter%3D22
50 Benjamin N Dykes, PhD, Astrology of the World, Volume I 2013
51 Jacques Gaffarel, Unheard-of curiosities: concerning the talismanical sculpture of the Persians, the horoscope of the Patriarkes; and the reading of the stars (trans. 1650)

Chapter 4
52 Ovid, Metamorphoses 13. 687 (trans Melville)
53 Pseudo-Hyginus, Astronomica 2.21
54 https://en.wikipedia.org/wiki/Aeneid

Chapter 5
55 Melanie Reinhart, Chiron and the Healing Journey, 2009
56 https://www.skyatnightmagazine.com/space-science/did-comets-bring-life-to-earth
57 DNA building blocks can be made in space By NASA's Goddard Space Flight Center, Greenbelt, Maryland https://www.astronomy.com/science/dna-building-blocks-can-be-made-in-space/
58 https://en.wikipedia.org/wiki/1910
59 https://en.wikipedia.org/wiki/Brain_(computer_virus)
60 https://en.wikipedia.org/wiki/Apple_Inc.
61 https://en.wikipedia.org/wiki/Steve_Jobs
62 https://en.wikipedia.org/wiki/IPhone
63 https://en.wikipedia.org/wiki/2012%E2%80%932013_Cypriot_financial_crisis
64 https://en.wikipedia.org/wiki/History_of_molecular_biology
65 Ovid, Metamorphoses 13. 687 (trans Melville)
66 https://en.wikipedia.org/wiki/Aeneid

Chapter 8
67 https://en.wikipedia.org/wiki/Halley%27s_Comet#1066
68 https://en.wikipedia.org/wiki/Emperor_Yingzong_of_Song
69 https://fr.wikipedia.org/wiki/1067
70 https://en.wikipedia.org/wiki/1067
71 Xu Zizhi Tongjian(續資治通鑑), https://ctext.org/wiki.pl?if=gb&chapter=758438&remap=gb
72 https://en.wikipedia.org/wiki/Tycho_Brahe
73 https://en.wikipedia.org/wiki/Great_Comet_of_1577
74 https://en.wikipedia.org/wiki/Great_Comet_of_1577

Notes & References

75. https://zh.wikisource.org/zh-hant/%E6%98%8E%E5%8F%B2/%E5%8D%B727 "萬曆五年十月戊子, 彗星見西南, 蒼白色, 長數丈, 氣成白虹。由尾、箕越斗、牛逼女, 經月而滅。
76. https://en.wikipedia.org/wiki/Bencao_Gangmu
77. https://no.wikipedia.org/wiki/Vitenskaps%C3%A5ret_1677
78. Merlini Anglici ephemeris or Astrological Judgements for the year 1678 by William Lilly
79. https://www.raremaps.com/gallery/detail/80292/cometa-ao-1677-voigt
80. https://en.wikipedia.org/wiki/Spermatozoon
81. https://en.wikipedia.org/wiki/1678_in_science
82. https://en.wikipedia.org/wiki/Russo-Turkish_War_(1676%E2%80%931681)
83. https://en.wikipedia.org/wiki/Statute_of_Frauds
84. https://en.wikipedia.org/wiki/C/2019_Y4_(ATLAS)
85. https://en.wikipedia.org/wiki/Rochdale_Society_of_Equitable_Pioneers
86. https://en.wikipedia.org/wiki/Texas_annexation
87. https://en.wikipedia.org/wiki/Mexican–American_War
88. https://en.wikipedia.org/wiki/1845
89. https://en.wikipedia.org/wiki/First_Conference_of_the_International_Woman_Suffrage_Alliance
90. https://en.wikipedia.org/wiki/1902
91. https://en.wikipedia.org/wiki/1902
92. https://en.wikipedia.org/wiki/Panic_of_1901#:~:text=The%20Panic%20of%201901%20was,of%20the%20Northern%20Pacific%20Railway.
93. https://en.wikipedia.org/wiki/1901
94. https://en.wikipedia.org/wiki/1902
95. https://en.wikipedia.org/wiki/1901
96. https://en.wikipedia.org/wiki/Alzheimer%27s_disease#History
97. https://en.wikipedia.org/wiki/1902
98. https://en.wikipedia.org/wiki/Teddy_bear
99. https://en.wikipedia.org/wiki/William_McKinley
100. https://en.wikipedia.org/wiki/Halley%27s_Comet
101. https://www.skyatnightmagazine.com/space-science/anti-comet-umbrellas-halley-inventions
102. https://en.wikipedia.org/wiki/Camille_Flammarion
103. https://en.wikipedia.org/wiki/Qing_dynasty
104. https://en.wikipedia.org/wiki/Manchurian_plague
105. https://en.wikipedia.org/wiki/Qing_dynasty
106. https://en.wikipedia.org/wiki/Puyi
107. https://en.wikipedia.org/wiki/Edward_VII
108. https://en.wikipedia.org/wiki/1910
109. https://en.wikipedia.org/wiki/Mark_Twain

110 https://en.wikipedia.org/wiki/1910
111 https://en.wikipedia.org/wiki/1911
112 https://en.wikipedia.org/wiki/1911
113 https://en.wikipedia.org/wiki/Hawley_Harvey_Crippen
114 ttps://en.wikipedia.org/wiki/1910
115 https://en.wikipedia.org/wiki/1966
116 https://en.wikipedia.org/wiki/1967
117 https://www.weather.gov/top/1966TopekaTornado
118 https://en.wikipedia.org/wiki/Opposition_to_United_States_involvement_in_the_Vietnam_War
119 https://en.wikipedia.org/wiki/Civil_rights_movement
120 https://en.wikipedia.org/wiki/1967_Hong_Kong_riots
121 https://en.wikipedia.org/wiki/Rhodesia
122 https://en.wikipedia.org/wiki/Thurgood_Marshall
123 https://en.wikipedia.org/wiki/World_Food_Programme
124 https://en.wikipedia.org/wiki/Pound_sterling
125 https://en.wikipedia.org/wiki/Antlia
126 https://www.cometography.com/lcomets/1995o1.html
127 https://en.wikipedia.org/wiki/Comet_Hale%E2%80%93Bopp
128 https://www.cometography.com/lcomets/1995o1.html
129 https://iopscience.iop.org/article/10.3847/PSJ/abd32c
130 https://en.wikipedia.org/wiki/Heaven%27s_Gate_(religious_group)
131 https://en.wikipedia.org/wiki/Diana,_Princess_of_Wales
132 https://en.wikipedia.org/wiki/J._K._Rowling
133 https://en.wikipedia.org/wiki/Euro
134 https://en.wikipedia.org/wiki/1998_Russian_financial_crisis
135 https://en.wikipedia.org/wiki/Influenza_A_virus_subtype_H5N1
136 https://en.wikipedia.org/wiki/Deep_Blue_(chess_computer)
137 https://www.elon.edu/u/imagining/time-capsule/150-years/back-1960-1990/
138 McNaught discovered the comet in a CCD image on 7 August 2006 during the course of routine observations for the Siding Spring Survey, which searched for Near-Earth Objects that might represent a collision threat to Earth.
139 https://en.wikipedia.org/wiki/Comet_McNaught
140 https://en.wikipedia.org/wiki/2013_Egyptian_coup_d%27état

www.ingramcontent.com/pod-product-compliance
Lightning Source LLC
Chambersburg PA
CBHW072129160426
43197CB00012B/2043